LENS ABERRATION DATA

Monographs on Applied Optics

No. 1

LENS ABERRATION DATA

J. M. PALMER, M.Sc.

Rank Precision Industries Ltd

WITH A PREFACE BY

W. D. WRIGHT

Professor of Applied Optics

Imperial College of Science and Technology

ADAM HILGER

LONDON

ISBN 0 85274 160 X

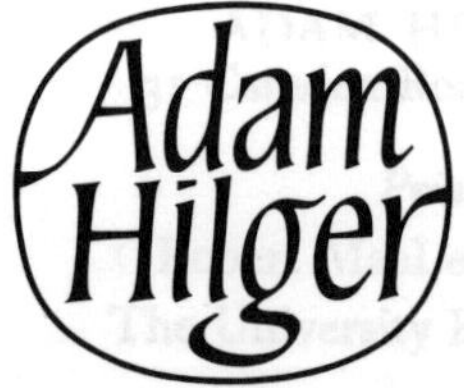

A company now owned by
The Institute of Physics
Techno House
Redcliffe Way
Bristol BS1 6NX
England

Preface

by

PROFESSOR W. D. WRIGHT

Applied Optics Section, Imperial College of Science and Technology

A year or so ago Mr Neville Goodman of Adam Hilger Ltd made the interesting suggestion that there might be some very publishable material among the M.Sc. reports prepared by our postgraduate optics students. These reports usually take the form of an intensive survey of some selected area of optics, perhaps with the addition of an original piece of analysis by the student. They have the advantage of being written by young and critical minds and they certainly deserve a wider readership than they can possibly achieve on the shelves of a college library. Even so, they are generally too specialized to call for more than limited editions.

This monograph by Mr J. M. Palmer is the first in the series which Mr Goodman is planning. It has developed from the project report written by Mr Palmer in 1968 and undoubtedly contains material of much interest and importance to both the designer and the user of lenses. I am very impressed indeed with Mr Palmer's keen insight into the problems of lens design and I particularly welcome his awareness of the historical development of the subject. As one of the generation of Professor Conrady's students who did his ray tracing with six-figure logarithms, I appreciate his remark on p. 35 that in those days 'the designer arranged his calculations in order that he could extract the greatest possible amount of information from every ray traced' and his tribute to Conrady that 'in this respect he was particularly notable in his contribution to the art of calculation'. Nowadays the computer has made us more prodigal with our data and we are in some manner like the old woman who lived in a shoe, having so many rays we hardly know what to do.

Lens design is, indeed, still an art and the successful designer must be the master and not the servant of his computer. I believe that in this book Mr Palmer shows very clearly how ray tracing should be directed to the required end and in doing so makes a most significant contribution to the advancement of the designer's art.

Author's Preface

The techniques of lens design have undergone considerable change since the days of Dolland (1706–1761), Fraunhofer (1787–1826) and Petzval (1807–1891), to whom we attribute some of the first corrected lenses. The work of Ernst Abbe (1840–1905) in developing and implementing the analytical methods of design produced considerable advances in instrument design, notably in microscope manufacture. Abbe recognized the earlier work of Seidel in the analysis of geometrical aberration. Seidel's treatment considered only linear and third-order terms but was adopted by Abbe to form a systematic method of design. The analytical approach to lens aberration has been developed, by more recent workers, to include high order coefficients. Buchdahl (1964–1968)* has given a comprehensive treatment.

Unfortunately, direct analytical methods are not very successful in solving design problems, so, for the analysis of aberration in a specific lens design, the techniques of geometrical optics must be used.

Geometrical optics is principally concerned with ray tracing whereby the stigmatic defects of rays may be determined. In Chapter 2, we shall see that the method involves applying a set of recurrence relationships to each surface taken from the object to the image space and then calculating the ray errors. Ray tracing also leads to another form of geometrical aberration—wave aberration, which is expressed as the departure from the ideal spherical wavefronts.

These calculations were originally performed using six-figure logarithms and were in consequence slow and tedious. Volume 1 of A. E. Conrady's *Applied Optics and Optical Design* (1929) gives worked examples of typical calculations.

The introduction of mechanical, and later electronic, calculators considerably speeded the process of ray tracing. A major advance followed—the invention of the high-speed electronic digital computer, which made available computing power unimaginable in Conrady's day. In parallel with the development of the techniques of geometrical optics, at least since the mid-forties, progress in optics has benefited greatly from the concept and methods of information theory applied to image formation theory (see Chapter 3).

This evolution in the understanding of optics was reflected in a change in

* BUCHDAHL, M.A., *Optical Aberation Coefficients with Collected Papers from J.O.S.A.* (Dover, 1968.)

fashion in the working methods of practising lens designers and instrument makers. The calculation and forms of analysis used by Conrady and his contemporaries were largely governed by the time involved in tracing rays. These workers overcame the difficulty with great ingenuity, using simple subsidiary calculations to extract the maximum information from each ray they traced so laboriously. The new computing power brought by the electronic computer tended to overshadow their methods. The geometrical merit function based mainly on transverse ray aberration became the centre of interest. The adoption of this somewhat crude measure of image quality lead to spectacular results in lens design. It enabled designers such as C. G. Wynne in Great Britain, D. Feder in the U.S.A., and E. Glatzel in Germany, to develop optimization programs involving iterative procedures to steer the design through multi-dimensional aberration space. The resulting lens designs were always orders of magnitude better in performance than those produced by manual methods. However, one must not single out the invention of the electronic computer for the entire credit. Other technological innovations have played their parts. For example, the introduction of single-layer anti-reflection coatings and of the highly efficient broad-band multi-layer coatings has enabled the designer to adopt more complex lens constructions. Also, the availability of a greater range of optical glass, much improved in quality, contributed to better lens performance. Freed from the earlier constructional limitations the electronic computer was able to demonstrate its vastly superior ability to handle a large number of design parameters in iterative lens design.

The stage has been reached where the problems of balancing aberration for optium performance, taking into account the diffraction effects of imagery, are being investigated by H. H. Hopkins in Great Britain, and W. King and A. Offner in the U.S.A. Their technique is mainly directed towards improving high performance systems designed by geometrical merit function methods. It has been known for some time that the lower performance systems can also benefit from diffraction-based criteria (see § 3.4).

With large high-speed electronic computers, comprehensive image evaluation is a realistic proposition. In retrospect, we might consider lens design development to have been accompanied by a remarkable collection of image analysis and aberration display techniques. It is tempting to dismiss many of these as mere whims of fashion or at least out-dated. A better informed approach might allow the view that such methods are worthy of examination.

Indeed, we find many of the earlier methods are still perfectly suited to certain design problems. Lens design is an expensive process and designers must be aware of the danger of over-powering the design problem with advanced techniques. We are no longer restricted by the extent of the calculation possible within the time available for design work; rather we have reached the position where the cost of computer time is to be considered as the limiting factor. It follows that judicious choice of the method of analysis leads to economic procedures.

A few of the forms of aberration analysis have been described in the scientific literature of optics. A great number, however, merely form the standard working methods of practising lens designers. The aim of this monograph is to survey the methods of lens aberration analysis with special reference to graphical presentation of the information. A linking theme which also forms the basis for critical comparison is provided in the form of numerical image analysis of a lens design of typical construction.

A considerable proportion of the material contained in this monograph was assembled whilst I was on a course of study at Imperial College of Science and Technology, University of London, preparing the dissertation I was required to submit for the M.Sc. degree. The extra material consists mainly of additional numerical data included for the sake of completeness and continuity of presentation.

A recurrent difficulty in any discussion of the mathematics of optics is the question of sign conventions. The one adopted here is that followed at Imperial College during my course of study. It is as follows: a set of Cartesian axes are constructed about the pole of the current surface so that dimensions measured to the right and upwards are positive. Angles measured clockwise from the optical axes are likewise positive. The exception to the sign convention occurs in paraxial ray tracing equations where the Conrady angle convection is used. In this case, a clockwise rotation of the axes onto the ray produces a positive sign and is thus the reverse of the Cartesian angle convention. A simple modification to the ray tracing equation will make them consistent if this should be required.

An attempt has been made to keep the nomenclature consistent but the varied sources of the material has made this difficult. In the main, I have adopted well-known symbols for quantities and tried to resist the temptation to invent my own. The convention for finite aperture quantities is to use upper case symbols, whereas the paraxial quantities are represented by the lower case.

The naming of rays is apt to cause confusion to those new to the subject of lens design. The term 'principal ray,' or in some countries the chief ray, refer to the ray which passes through the centre of the stop (crossing the optical axis). This ray is used to define field sizes. The image or marginal ray is that ray which crossed the optical axis at the object and image and passes through the edge of the stop. It thus defines numerical aperture.

I should like to acknowledge the generous advice and assistance given by Professor C. G. Wynne and the members of the Optical Design Group at Imperial College. Much of the numerical analysis could not have been completed without the use of computer programs and the programming experience of the Design Group. My thanks are due to Dr W. T. Welford for introducing me to many of the topics discussed in this monograph, for his sound and practical teaching, and for his valuable comments in the preparation of the dissertation. No less is my debt to Mr David Demaine of Rank Precision Industries Ltd, Analytical Division, Hilger & Watts for his unstinted encouragement both in furthering my studies in optics and in the preparation of this monograph. In particular, I make acknowledgement to him for the ray generation method described in § 2.2.2, which he developed to form the basis of ray generation in a comprehensive ray tracing and optimization program used for many years at Hilger & Watts to design a wide range of optical systems.

I am grateful for the assistance given me by Herr Martin Krautter of Carl Zeiss Oberkochen, who drew my attention to the work of Hilderbrand and described some of his own work in the study of diapoint analysis.

Thanks are due to Mr Neville Goodman and Mr David Tomlinson of Adam Hilger Ltd for their encouragement and assistance in the preparation of this monograph.

I am grateful to the following people for granting me permission to reproduced illustrations: Dr R. Kingslake (Figs. 14 and 15), the Editor of *Applied Optics* (Fig. 2), The Editor of the *Journal of the Optical Society of America* (Fig. 27), and the Editor of *Optica Acta* (Fig. 25).

J. M. PALMER

LONDON
August, 1970

Contents

1

Introduction

W. T. Welford has said, 'The physical principles underlying the design of optical instruments are very few in number, and yet the process of design has developed to a high level of sophistication and precision.' The aim of this monograph is to discuss some of the basic tools of this process. In the design of optical systems we are concerned, in the first place, with the linear properties of systems governed by the so-called paraxial or Gaussian optics. This aspect, however, turns out to be comparatively simple, and it is the second stage, that concerned with the non-linear properties, which requires the real design effort. This departure from Gaussian optics is described by the so-called aberrations of a system.

Traditionally, optical design has been concerned with geometrical optics, that is to say, using the formulae obtained from the electromagnetic theory of Maxwell, when the limiting process of allowing the wavelength to tend to zero has been applied. (See for example Born and Wolf, 1964.)

The most important of the physical principles involved in geometrical optics are Fermat's principle and Snell's law. These can be shown to be substantially equivalent, that is, the whole of geometrical optics can be deduced from either theorem. From Snell's law we can develop the concept of a 'ray of light', and hence ray tracing and ray aberration; while from Fermat's principle we develop the concept of wavefronts and wavefront aberration. Welford (1967) has given a most helpful diagram (Fig. 1) showing the relationship between these various ideas.

Of recent years, a considerable interest has been shown in the application of scalar wave theory, that is, diffraction theory, to the field of optical design. This interest has by no means resulted in the ousting of geometrical optics from its long-held position, but

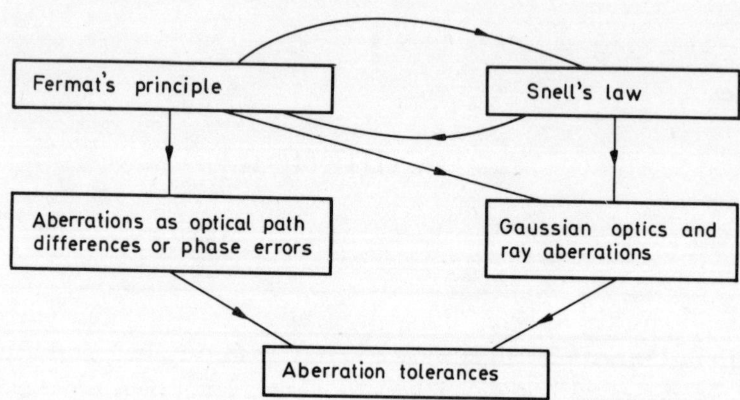

FIG. 1. The relationship between Fermat's principle, Snell's law and aberration tolerances.

it has, nevertheless, had a significant impact upon the subject. Diffraction theory is used mainly in assigning aberration tolerances, but also attempts have been made to apply it to the process of optimization of optical systems (Hopkins,* 1966).

Before attempting to deal with the various methods of approach to aberrations of optical systems, it is perhaps profitable to consider the reasons for wishing to display aberration data. These can be divided into three groups; the aberration data can be examined in terms of:

1. An aid to design.
2. Specification of a design and its suitability for inclusion in an optical system.
3. Assessing the manufacturing quality.

Let us look at these in turn.

The first is the most immediately obvious application of aberration data display, for it is the aim of the designer to minimize the aberrations to be within some limit in order to obtain the best possible performance for a given specification. The designer will require to keep a check on the progress of design, to know when to apply the correction emphasis appropriate to the aspects of the aberration most damaging to performance, and finally to decide when the design has converged to the optimum solution.

* When I refer in the text to Hopkins I am referring to H. H. Hopkins.

In the second of these, it is convenient to introduce the engineering term 'tolerance' to describe the analysis of aberration data in a form that tells us whether a design meets the specification for its intended application.

In many cases the optical system is only part of a larger electro-optical system involving electronic and mechanical components. In order to assess the performance of the overall system, and to examine the matching of the optical and electronic systems, it is useful to be able to present the aberrations as a frequency response.

Lastly, once the objective has been manufactured, it must be tested or subjected to some form of quality control. That is, the performance of the assembled optic must be compared to the predicated performance of the design. The aberration must therefore be presented in a form that can be compared with measurable quantities.

It is proposed to examine the methods of calculating and analysing aberration in relation to these requirements. Also, in order to correlate the properties of various methods of presentation, a numerical example has been selected and its aberration evaluated. The actual choice of design was largely arbitrary, consistent with being an optimized design of not necessarily diffraction limited performance, but being representative of a class of objective typically met in optical design. In fact, the design used is one due to Kidger and Wynne (1967) and is shown in Fig. 2. It falls in the group of systems classified by the term

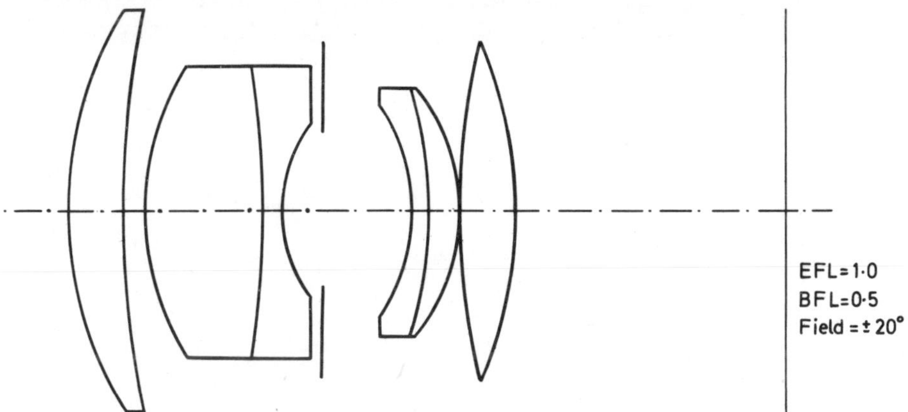

FIG. 2. A sectional drawing of the double Gauss objective. (After Kidger and Wynne, 1967.)

'double Gauss', and is optimized at an aperture ratio of $f/2$ and a semi-field angle of 20°; there is vignetting of about 65 per cent at the 20° field angle. The chromatic aberration was balanced for the C–F wavelength range using the glass type listed in the Chance–Pilkington catalogue (1966). The image analysis was consequently based on ray tracing calculations at three wavelengths, the F-line of helium at 486·1 nm, the hydrogen d-line at 587·5 nm and the helium C-line at 656·3 nm. The objective was optimized for an infinite object conjugate and a back focal length of 0·5 of the effective focal length. The system data given in Table 1.1 relate to a unit focal length.

TABLE 1.1

Surface number	Curvature	Separation	Aperture	n_d	v	Glass type
1	1·3764		0·60			
		0·0941		1·70000	41·18	BF 700412
2	0·2859		0·60			
		0·0472		1·0		
3	2·2460		0·44			
		0·2111		1·62252	60·32	DBC 623603
4	−0·2620		0·44			
		0·0300		1·74842	27·85	DEDF 748278
5	3·4205		0·44			
		0·0584		1·0		
6	0*		0·292			
		0·1781		1·0		
7	−3·6163		0·52			
		0·0300		1·57433	42·62	LF 575426
8	−1·1260		0·52			
		0·0641		1·69100	54·80	SBCA 691548
9	−2·5917		0·52			
		0·0000		1·0		
10	0·5897		0·72			
		0·1066		1·69100	54·80	SBCA 691548
11	−1·3418		0·72			

* Aperture stop.

2

The Aberrations of Geometrical Optics

Geometrical optics, we have seen, can be developed from Snell's law and Fermat's principle. These theorems lead respectively to the concepts of 'rays' and 'wavefronts'. Rays can be thought of as loci along which light energy is transported, and wavefronts as surfaces orthogonal to rays originating at a single point. The aberrations of geometrical optics are then defined from the ray errors obtained by comparing finite aperture rays to ideal rays; and the wavefront errors are obtained by comparing the wavefront to a spherical surface (the reference sphere).

Ray aberrations may take any one of several forms, such as longitudinal ray aberration and transverse ray aberration. Additionally, we can obtain some specialized forms of analysis from ray errors. The wavefront aberration is normally expressed in terms of the Conrady optical path differences (OPD) measured in units of wavelength.

All of the aberrations of geometrical optics are determined from ray tracing finite aperture rays with the equations developed from Snell's law.

2.1. RAY TRACING

Ray tracing is fundamental to the process of optical design and so warrants brief mention. In essence, it is the successive application of Snell's law for an infinitely narrow pencil of light (a ray) travelling from the object space to the image space. The resulting set of recurrence relationships, when presented in their full generality for a skew ray passing through a non-rotationally symmetrical set of surfaces, bears little superficial resemblance to the familiar form of Snell's law:

$$n \sin I = n' \sin I'$$

where n is the refractive index in the object space, n' the refractive index in the image space, I the angle of incidence, and I' the angle of refraction. Indeed, there can be many formulations of these relations for the general case and even for special cases such as a ray traced in the meridian section. Welford (1956) gives a bibliography of these methods. In this monograph the image analysis graphs for the double Gauss example were prepared from ray tracing data obtained using an algebraic ray tracing method due originally to Max Lange in 1909, reproduced by J. P. C. Southall (1913), and given by Welford (1963a) in a form adapted for use with calculating machines and electronic computers.

Snell's law can be generalized to a vector form:

$$n\mathbf{r}_\wedge \mathbf{a} = n'\mathbf{r}'_\wedge \mathbf{a}$$

where $\mathbf{a}(\alpha, \beta, \gamma)$ is the unit vector normal to the surface, $\mathbf{r}(L, M, N)$ the unit vector along the ray incident to the surface, and $\mathbf{r}'(L', M', N')$ the unit vector along the ray refracted by the surface. The terms in parenthesis are the direction cosines of the vectors. Hence with a standard vector expansion we obtain

$$n'\mathbf{r}' - n\mathbf{r} = (n'\mathbf{r}' . \mathbf{a} - n\mathbf{r} . \mathbf{a})\mathbf{a} = (n' \cos I' - n \cos I)\mathbf{a}$$

The form of the recurrence relationships for tracing skew rays through spherical surfaces can be divided into two parts:

1. The transfer equations.
2. The refraction equations.

1. Let the co-ordinates of the point P_{-1} in the co-ordinate system of the previous surface be (X_{-1}, Y_{-1}, Z_{-1}) and let those of P_0 (in the current co-ordinate system) be $(\underline{X}, \underline{Y}, 0)$, as in Fig. 3. Further, let the separation of the two systems be distance d and the curvature of the current surface be c. Then,

$$\underline{X} = X_{-1} + \frac{L}{N}(d - Z_{-1})$$

$$\underline{Y} = Y_{-1} + \frac{M}{N}(d - Z_{-1})$$

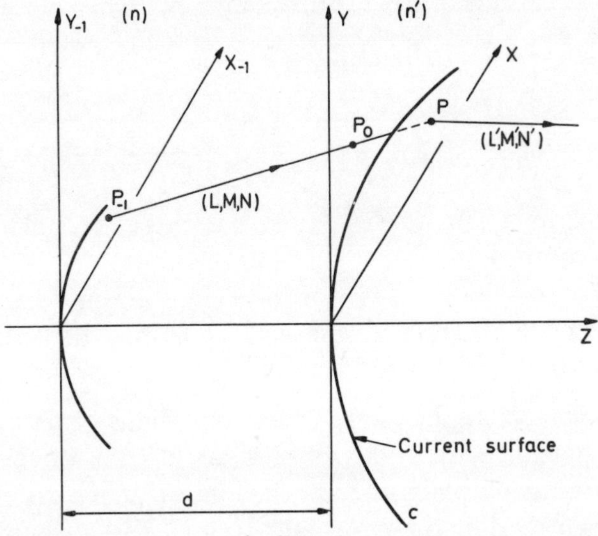

FIG. 3. Transfer and refraction in skew ray tracing.

at the tangent plane to the current surface. To complete the transfer to the ray interception $P(X, Y, Z)$ with the current surface we introduce the new variables F, G and Δ, where Δ is actually the segment PP_0. We have

$$F = c(\underline{X}^2 + \underline{Y}^2)$$

$$G = N - c(L\underline{X} + M\underline{Y})$$

$$\Delta = \frac{F}{G + (G^2 - cF)^{\frac{1}{2}}}$$

Substituting

$$X = \underline{X} + L\Delta$$

$$Y = \underline{Y} + M\Delta$$

$$Z = N\Delta$$

gives the co-ordinates of P.

2. For refraction at P, we use the generalized power K of the surface, where $K = c(n' \cos I' - n \cos I)$, and can show

$$\cos I = +(G^2 - cF)^{\frac{1}{2}}$$

and

$$n' \cos I' = (n'^2 - n^2 + n^2 \cos^2 I)^{\frac{1}{2}}$$

Then the direction cosines are given by

$$n'L' = nL - XK$$

$$n'M' = nM - YK$$

$$n'N' = nN - ZK + n' \cos I' - n \cos I$$

Check equations can be developed but these are not used with electronic computers.

Also, we may note that, for rays in the meridian section, $L = X = 0$ and consequently the equations may be reduced to a simplified form.

The successive application of these equations generates the co-ordinates and direction cosines of a ray at each of the surfaces and their tangent planes from the object space to the image space, thus providing the basic data for image assessment.

In the Gaussian region, rays are traced by paraxial ray equations which can be regarded as a special case of the above equations for limitingly small apertures when we allow $\sin I \to I$ and consequently $Y \to h$, $M/N \to u$ and $Z_{-1} \to 0$.

The transfer equation is

$$h_{+1} = h - d'u'$$

and the refraction equation

$$n'u' = nu + hk$$

at a surface where the paraxial ray height is h and the paraxial convergence angles of the ray are u and u' in the object and image spaces of a surface respectively. The power k of the surface becomes

$$k = c(n' - n)$$

The equations can be developed to deal with ray tracing through aspheric surfaces, given, for example, by Smith (1945) and Welford (1952).

2.2 RAY GENERATION

The advent of fast electronic computers of large storage capacity made possible the tracing of large numbers of rays and comprehensive image analysis in a short time at a reasonable cost.

A skilled designer can make an adequate assessment of the performance of a system from a small number of rays and so rarely needs to trace more than ten to fifty rays. In this case it is generally possible to supply the initial ray data at the object in the form of a list of co-ordinates and direction cosines. However, even for fifty rays, the task of preparing the list of data becomes time consuming and subject to human error. It is, therefore, natural to turn to methods of specifiying rays in the form of a list of a dozen or so numbers which allow the computer to calculate all the initial ray data it requires to trace a set of rays.

The degree of sophistication of this technique depends on the computer available and the task for which the ray tracing is intended. Therefore, a general survey cannot be attempted, but three possible methods from the author's experience will be described.

All three methods need to be supplied with data defining the Gaussian parameters of the system. These are generally given by the transverse magnification, the numerical aperture, and the stop position. Alternatively, the latter quantity may be implied by defining the paraxial image (marginal) ray and the principal ray co-ordinates and convergence angles at some reference plane. In either case the application of paraxial ray tracing will generate the co-ordinates and angles of the paraxial image and principal ray from the object to image spaces, including the entrance pupil, stop and exit pupil data.

A fairly simple approach is taken as the first example. We produce a list which consists of three numbers per ray:

1. A label to identify the ray, indicating whether it is a paraxial, meridian or skew ray.
2. A relative pupil height indicating at what fraction of the maximum pupil height we require the ray to be traced. For a skew ray we shall need an additional quantity. We can supply the relative pupil height and an azimuth angle, or, alternatively, two relative pupil heights, one for the x co-ordinate direction and one for the y co-ordinate direction.
3. The object heights at which we require the corresponding image analysis or perhaps simply the angles of the principal rays at the entrance pupil defining these image points.

For the sake of convenience we may call this form of ray generation relative ray data.

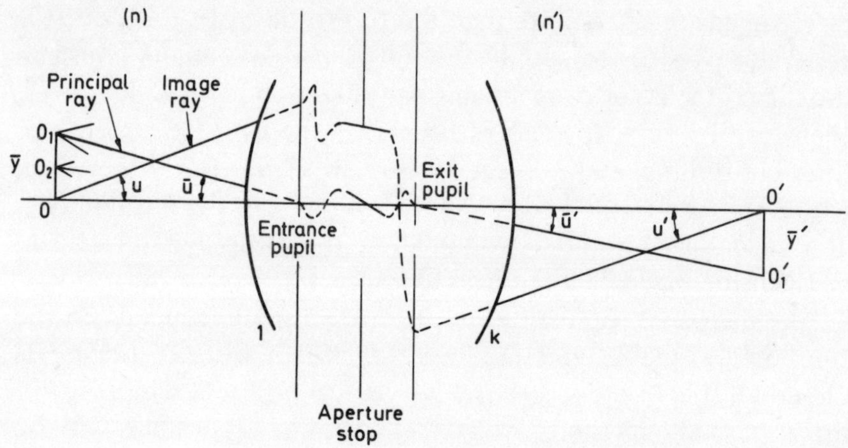

FIG. 4. The pupil defining paraxial rays used for the generation of fans of finite aperture rays.

2.2.1. *Relative ray data*

Suppose, as in Fig. 4, that the Gaussian parameters are defined by the following variables:

Principal ray data at the object (or first surface for an infinite object conjugate)

Paraxial convergence angle $\bar{u}$
Object height $\bar{y}$

Image (*marginal*) *ray data at the object* (or first surface for an infinite object conjugate)

Paraxial convergence angle u
Ray height y (zero for finite conjugate)

This initial data set is followed by the ray list. The first number is an integer identifying the ray by the convention

1 A paraxial ray.
2 A finite meridian ray.
3 A skew ray.

The ray tracing program may then select the appropriate subroutine to give the most efficient form of tracing. The pupil data set for a meridian ray is defined by a relative pupil height b, where

$$-1{\cdot}0 < b < 1{\cdot}0$$

Skew ray pupil data may be defined by two relative pupil heights:

Relative ray height in the x-direction (sagittal section) a, where

$$-1{\cdot}0<a<1{\cdot}0$$

Relative ray height in the y-direction (tangential or meridan section) b, where

$$-1{\cdot}0<b<1{\cdot}0$$

Object Data. The ray heights of intermediate points at the object are $c\bar{y}$, where

$$-1{\cdot}0<c<1{\cdot}0$$

or if the field angle θ is given, then $c\bar{y}$ must be calculated.

The program generates the ray data at the object or first surface using

$$R=[1+(c\,.\,\bar{u}+b\,.\,u)^2+(a\,.\,u)^2]^{\frac{1}{2}}$$

$$L=\frac{a\,.\,u}{R}$$

$$M=\frac{c\,.\,\bar{u}+b\,.\,u}{R}$$

$$Y=c\,.\,\bar{y}$$

Thus, for a given field angle defined by the value of c or θ, we can generate a fan of rays across the pupil by taking a series of values of a and b between $+1{\cdot}0$ and $-1{\cdot}0$.

2.2.2. *Automatic generation of relative ray data*

This is another approach which is very simple from the point of view of data input, but requires rather more computer store than the previous method. We are supplied with the same Gaussian data, but, for skew rays, the densities of rays in the entrance window (object plane usually) and the entrance pupil are given for the meridian and sagittal sections.

In some non-rotationally symmetrical systems such as spectrographs, it may be necessary to define the principal ray and image ray in both the meridian and sagittal sections. This general case will be described. Suppose at some reference plane, as in Fig. 5, we define:

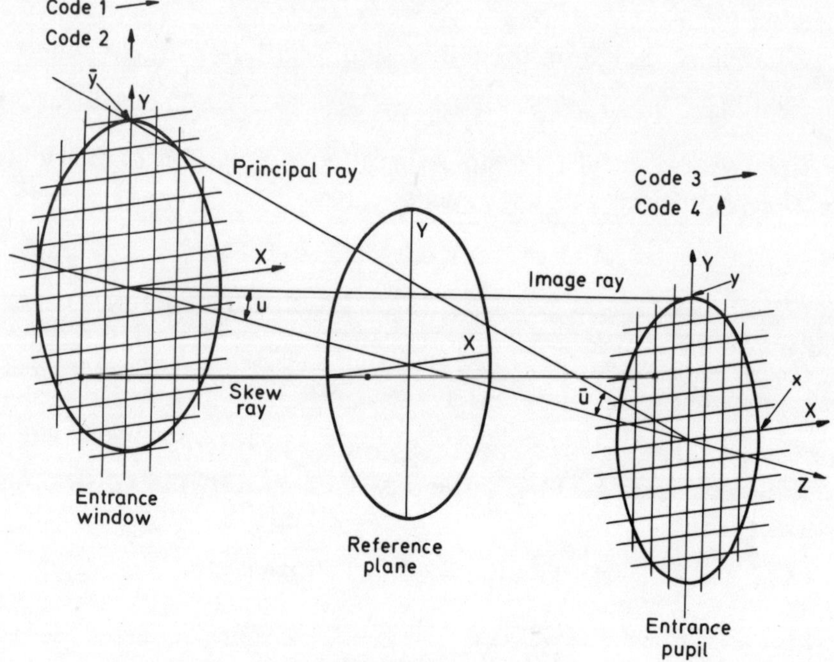

FIG. 5. The pupil meshes used for the automatic generation of finite aperture rays.

	X-direction	*Y*-direction
Principal ray data		
Convergence angles	$\bar{v}$	$\bar{u}$
Ray height	$\bar{x}$	$\bar{y}$
Image ray data		
Convergence angles	v	u
Ray height	x	y
Entrance window densities	IDENWX	IDENWY
Entrance pupil densities	IDENPX	IDENPY

The position and height of the object are effectively defined by this data set. The position of the object can be obtained by finding the intersection of the optical axis with the paraxial marginal ray in the object space. The height of the paraxial principal ray in the focal plane then corresponds to the object height. Similarly, the paraxial principal and image rays in the image space define the image position and height. Alternatively, if the front conjugate, stop position, numerical aperture, and magnification are given then the principal and image rays can be calculated. The computer now sets up matrix-like arrays representing grids in the entrance window and pupils. The ray positions in the grids are referenced by means of codes derived from an $X-Y$ rectangular co-ordinate system in the window and pupil. We generally assume that the grid in the entrance pupil is in geometrical correspondence to the exit pupil grid; thus we can effectively ignore the pupil aberration. This may not always be justified as, for example, in wide-angle objectives based on the Aviogon configuration, when image analysis must be referred to the entrance pupil heights rather than to the exit pupil heights. Stop aberration and image illumination are then treated separately.

The codes are normalized to unity so that, for example, a circular pupil will have a unit radius and the sum of the squares of the codes of any transmitted skew ray will be less than unity. Further, we may obtain a geometric series of points in the entrance window by taking the square root of CODE 1 and CODE 2. The codes are calculated as follows:

$$\text{CODE 1} = \frac{2(\text{IWX}-1)}{\text{IDENWX}-1} - 1{\cdot}0 \qquad \text{IWX} = 1 \text{ to IDENWX}$$

$$\text{CODE 2} = 1{\cdot}0 - \frac{2(\text{IWY}-1)}{\text{IDENWY}-1} \qquad \text{IWY} = 1 \text{ to IDENWY}$$

$$\text{CODE 3} = \frac{2(\text{IPX}-1)}{\text{IDENPX}-1} - 1{\cdot}0 \qquad \text{IPX} = 1 \text{ to IDENPX}$$

$$\text{CODE 4} = 1{\cdot}0 - \frac{2(\text{IPY}-1)}{\text{IDENPY}-1} \qquad \text{IPY} = 1 \text{ to IDENPY}$$

These equations have been written in terms of computer program variables. The letter I indicates an integer; the letter P or W indicates the entrance pupil or entrance window; X and Y have their usual meaning and IDEN indicates the density of points across the pupil or window. The variable names are thus built up according to the appropriate significance, as for example IDENWY is the integer representing the number of points in the entrance window taken in the Y-direction.

It follows that a principal ray will have CODE 3 = CODE 4 = 0 and a meridian ray will have CODE 1 = CODE 3 = 0. Once the codes for each of the values of the running variables, IWX, IWY, IPX, IPY, have been evaluated any ray can be defined completely at the reference plane by its codes and the Gaussian ray data of the reference plane given above.

In order to carry out the ray trace the codes must be converted to ray trace parameters:

$$X = \bar{x}\ \text{CODE}\ 1 + x\ \text{CODE}\ 3$$

$$Y = \bar{y}\ \text{CODE}\ 2 + y\ \text{CODE}\ 4$$

$$L = \frac{\bar{v}\ \text{CODE}\ 1 + v\ \text{CODE}\ 3}{R}$$

$$M = \frac{\bar{u}\ \text{CODE}\ 2 + u\ \text{CODE}\ 4}{\text{R}}$$

$$N = \frac{1}{R}$$

$$R = [1 + (\bar{v}\ \text{CODE}\ 1 + v\ \text{CODE}\ 3)^2 + (\bar{u}\ \text{CODE}\ 2 + u\ \text{CODE}\ 4)^2]^{\frac{1}{2}}$$

The first operation is to transfer from the reference plane to the object plane and then the computer can proceed with the ray trace as before, compiling a list of data in a grid at some position in the image space, usually the exit window. This list is suitable for image analysis.

In many cases, owing to the symmetry of the system, it is unnecessary to trace all the rays. To obtain the complete list, it is only necessary to equate the data for the set of pupil codes having the same absolute value about the axis of symmetry. By this means we can achieve a considerable saving of computer time.

2.2.3. *Automatic generation of aperture filling ray fans*

This last form of automatic ray generation is of most interest when dealing with the valuable optical design technique known as vignetting.

In many design problems the marginal rays from extreme field points have so much aberration associated with them that they do not usefully contribute to good image formation but merely add to the stray light or flare in the image. In these cases it is usual to reduce the diameter of one or more of the lens components, thereby obstructing the passage of the offending rays, whilst allowing a full axial fan of rays through the objective.

We can more easily investigate the effect of assigning diameters to each of the objective's surfaces if the computer automatically generates the limiting fan of rays in each section for the various field angles. The procedure will be to trace a full aperture meridian ray from an axial point and to check that the specified diameters of each surface allow sufficient clear aperture to transmit the ray. Vignetting of the axial full aperture ray implies that the required numerical aperture cannot be obtained. Next, the principal ray at each of the required field angles will be traced and a similar check made. Vignetting of the principal ray is generally not permitted. Then a series of meridian rays of progressively increasing aperture are traced on both sides of the principal ray until the edges of the unvignetted aperture are found. The ray midway between the extreme rays is found and used as a new reference ray. The meridian rays at the 0·7 zone are traced to complete the meridian section data.

Skew rays relative to the new reference ray are traced at the full unvignetted aperture and 0·7 zone for the sagittal section and the 45° azimuth. Fig. 6 illustrates the ray intersection with a vignetted pupil. It is assumed that symmetry about the meridian plane exists, so that only a semi-pupil need be traced.

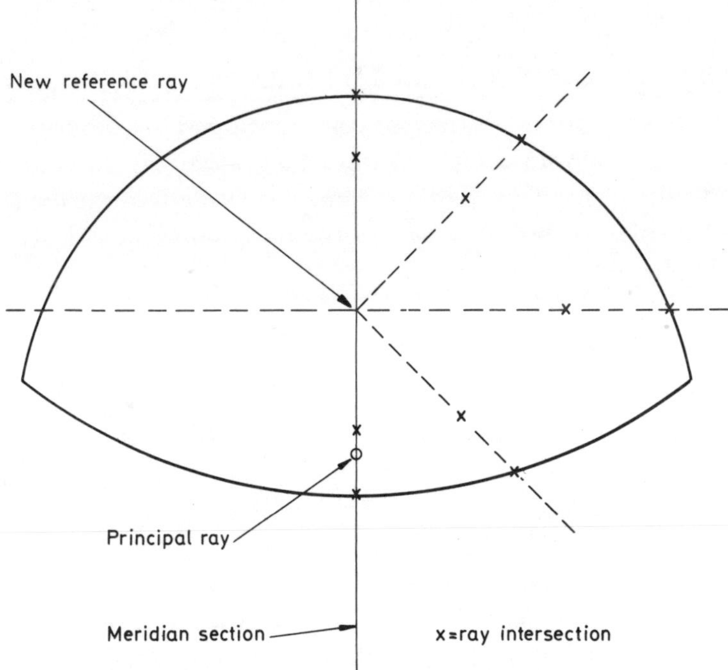

FIG. 6. The intersection of rays in the pupil for a pupil-filling ray generation procedure.

2.3. IMAGE ANALYSIS—TOTAL RAY ABERRATIONS

In § 2.1 we saw that Snell's law could be adapted to give a set of ray trace recurrence relationships and in § 2.2 considered some methods of applying these relationships to a system to produce the co-ordinates and direction cosines of rays in the image space referred to the chosen image plane. However, in this form it is virtually impossible to make an appraisal of the information in terms of its image forming qualities. In the context of ray generation it was noted that the fast electronic computer makes possible the tracing of large numbers of rays, but now we can see this very quality implies that further processing of the data is imperative for the rapid assimilation of the information in a readily comprehensive form. Consequently, the problem of image analysis has become of considerable interest.

Now, while for the non-physicist the formation of an image by an optical system remains somewhat of a mystical process, even for the physicist the phenomenon is not without its difficulties. Foremost amongst these is the difficulty of defining the image in such a way that critical analysis is possible. There is also the problem of reducing the three dimensional properties of the image to readily comprehensive terms. It is not surprising, therefore, that numerous methods have been developed to describe and display the image forming qualities of a system. In the following section and subsequent ones an attempt is made to describe some of the common techniques used in the design and specification of optical systems.

2.3.1. *Longitudinal ray aberration*

The traditional method of displaying lens aberration data is to plot the longitudinal ray aberration. At first sight, at least, this is the obvious way of describing the aberration as it corresponds to the distance through which a ground glass screen must be moved to change from an axial ray focus to a marginal ray focus in the presence of the simplest form of aberration, namely, spherical aberration. Also, it is obtained directly from a trigonometrical ray trace calculation of the type described by Conrady (1929). It has the disadvantage that it may attain any value between plus infinity and minus infinity without necessarily being of very serious importance in its effect on the final image; in this respect it can be a very misleading and inconvenient measure of aberration.

In Fig. 7 is shown longitudinal ray aberration for an extra-axial image point formed by the intersection of the principal ray

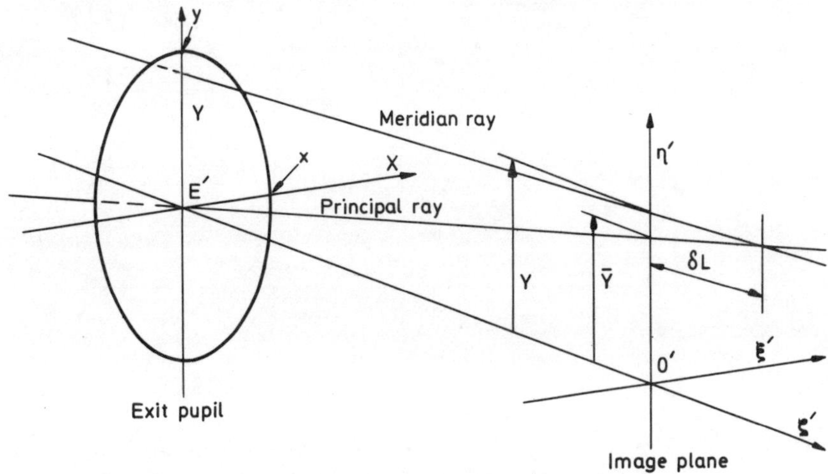

FIG. 7. Longitudinal ray aberration in the image space.

and a meridian ray from an intermediate zone in the pupil. The longitudinal aberration is measured parallel to the optical axis from the Gaussian image plane to the plane through this point of intersection. It is only in the meridian section that longitudinal aberration has any significance, since it is possible that a ray traced skew to this plane will never intercept the principal ray.

Referring to Fig. 8 we see that in the image plane the ray intersections are P and $\bar{\text{P}}$ having the separation $Y-\bar{Y}$ and in the intersection plane the corresponding distance is $\delta L(\tan\bar{\theta}-\tan\theta)$. Hence equating $Y-\bar{Y}=\delta L(\tan\bar{\theta}-\tan\theta)$ we obtain

$$\delta L=\frac{Y-\bar{Y}}{(\bar{M}/\bar{N})-(M/N)}$$

It follows that overcorrected spherical aberration will have a positive sign. The effect of a curved image plane can be included in the aberration equation. In this case, the final operation of the ray trace is to transfer to the curved plane, thereby determining the distances Z and $\bar{Z}$ from the tangent plane to the intersection

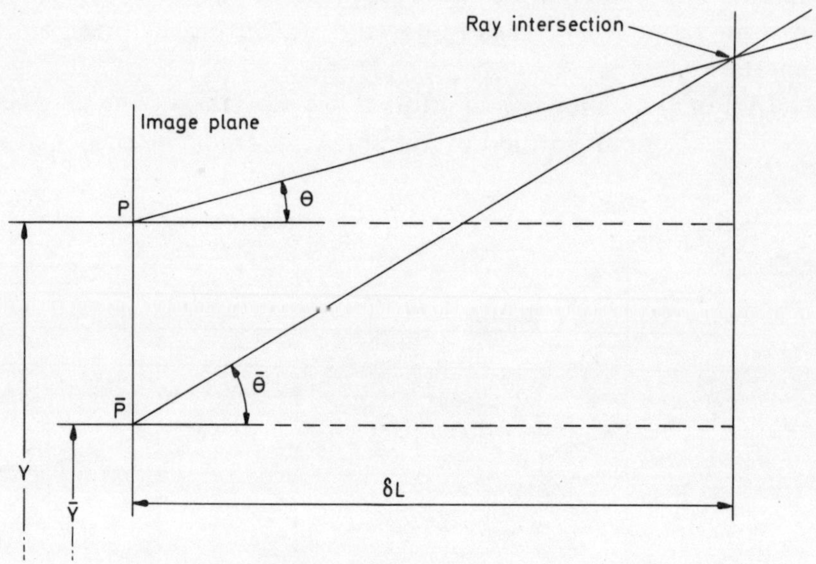

FIG. 8. Longitudinal ray aberration—the meridian section in the focal region.

with the current image surface for the meridian ray and principal ray respectively. Then the longitudinal aberration is given by

$$\delta L = \frac{Y - \bar{Y} + (\bar{Z} - Z)(M/N)}{(\bar{M}/\bar{N}) - (M/N)}$$

As a graphic display the normalized pupil height H is plotted as ordinate against longitudinal ray aberration as abscissa. Each field angle is given a separate Cartesian system and, for the sake of clarity, the longitudinal aberration curves for the various wavelengths may be shown in colours representing their wavelengths, or, as in Fig. 9, the principal wavelength as a continuous line and the extreme wavelengths as broken lines. It is usual to define a single image plane for all wavelengths and to select the Gaussian image plane of the principal wavelength for this purpose. As the effect of defocus is represented by the shifting of the ordinate axes to the left or right, the choice of image plane is not critical and has no influence on the shape of the curves.

2.3.2. *Transverse ray aberration*

Another effect of the application of electronic computers to optical design was the almost universal adoption of transverse

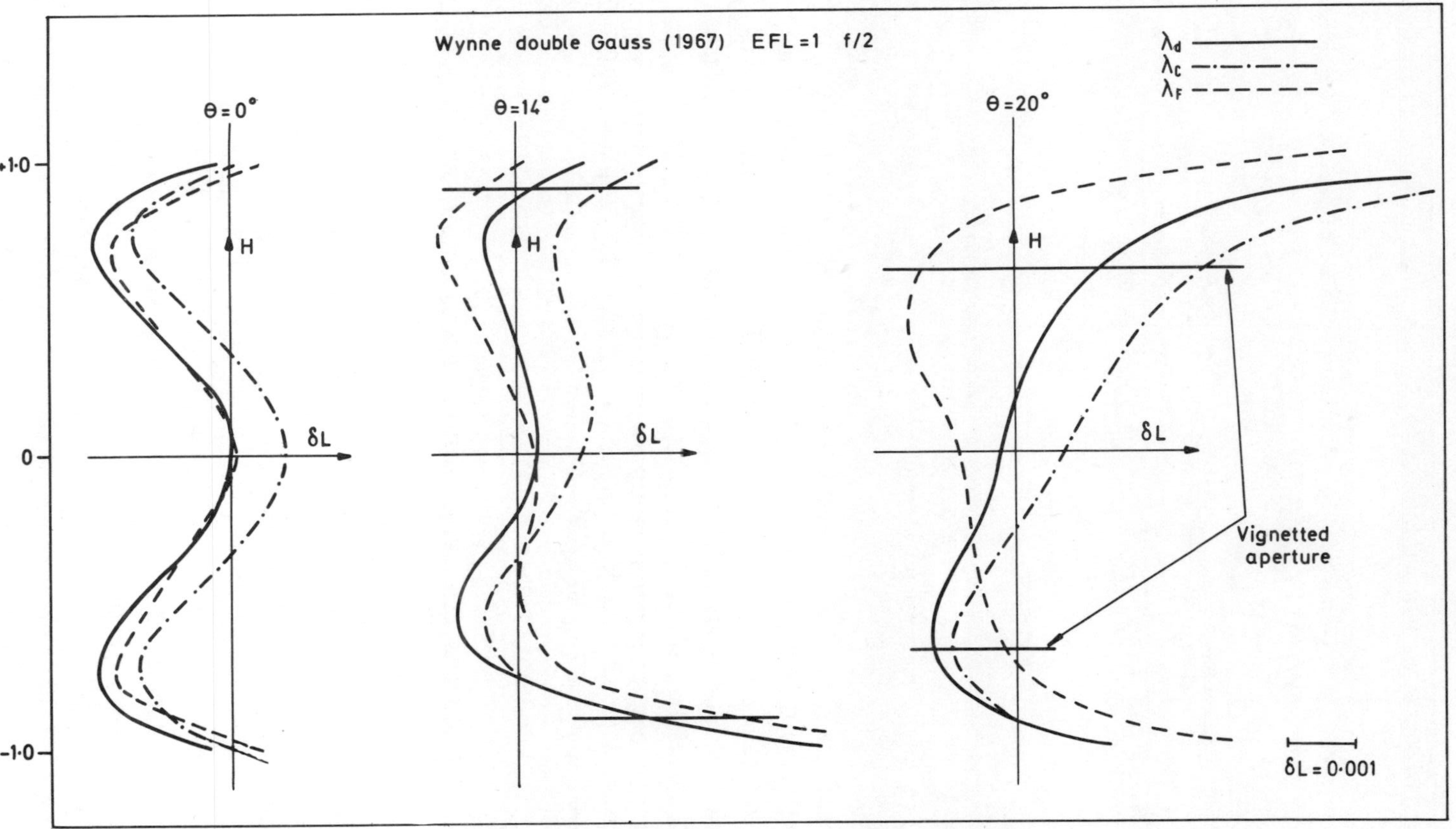

FIG. 9. Longitudinal ray aberration plotted for the Wynne double Gauss design.

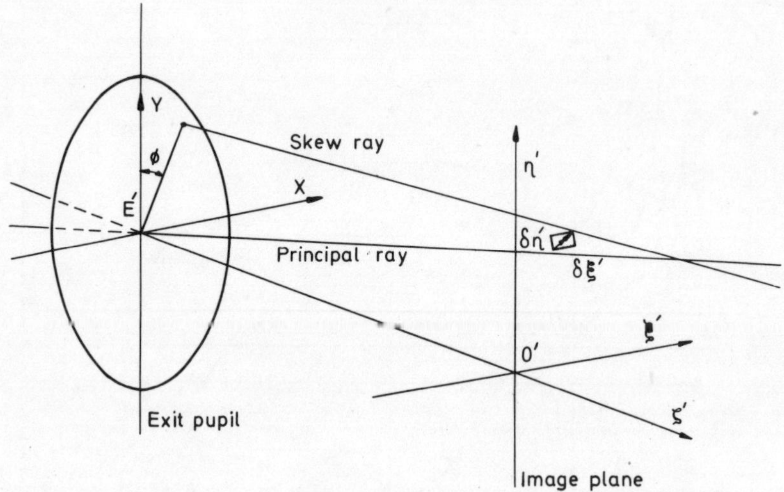

FIG. 10(*a*). Transverse ray aberration of a skew ray in the image space.

aberration. There were several reasons for this change but it is worth mentioning that the transverse error indicates the image spread of a point object and has a simple relationship to wave-front aberration. In addition there is no restriction to meridian rays, as can be seen in Fig. 10(*a*).

Transverse aberration in the meridian section is usually considered to be of most interest. It is calculated from the ray trace data in one of two ways:

1. The total transverse ray error $\delta\eta' = Y - y$, where y is the paraxial height.
2. The transverse error with respect to the principal ray of the principal wavelength, thus eliminating the effect of distortion, gives $\delta\eta' = Y - \bar{Y}$.

The ray height Y is taken in the chosen image plane, usually the Gaussian image plane of the principal wavelength. The sign of the aberration will be such that, for a positive-going image, a positive error indicates that the image ray is above the principal ray. For the double Gauss example the second method was employed in calculating the transverse aberration.

In the case of skew rays [Fig. 10(*b*)] the transverse aberration is considered to have two perpendicular components $\delta\xi'$ and $\delta\eta'$ giving a resultant error of $s = (\delta\xi'^2 + \delta\eta'^2)^{\frac{1}{2}}$, which may be treated as a radius vector from the true image point. If we assume that

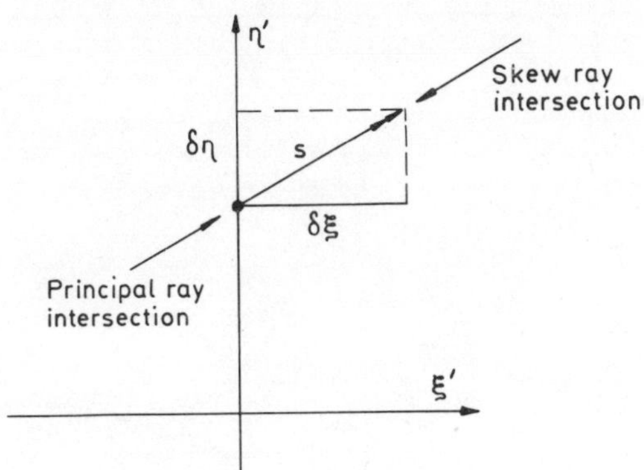

FIG. 10(*b*). The components of transverse ray aberration in the image plane.

the principal ray is in the $X-Y$ plane as meridian section as before, again we may calculate the error s in one of two ways—with respect to either the principal ray or the Gaussian image point.

Graphically we display the meridian transverse aberration by plotting the normalized pupil height H as ordinate against transverse ray aberration $\delta\eta'$ as abscissa. R. E. Hopkins (1967) chooses to reverse these axes but this does not seem to be general practice and so has not been adopted in the aberration plots shown in Fig. 11. His convention, however, does have the advantage of clearly differentiating between longitudinal and transverse aberration and can be adopted for this end.

The form of the skew ray aberration is usually examined by tracing rays from object points in the meridian section through pupil points in the sagittal section and perhaps the 45° azimuth, and then plotting the two components $\delta\xi'$ and $\delta\eta'$ as abscissa against the normalized pupil height. Sometimes the $\delta\eta'$ plot is omitted as this component is usually very much smaller than the $d\xi'$ component.

The display of the vector error is discussed in §§ 2.3.5 and 2.3.6. The relationship between transverse ray aberration and wavefront aberration will be considered more fully in § 2.5, but it is useful to note here that the meridian transverse error is proportional to the first differential coefficient of the wavefront aberration function [see Hopkins (1950), Chapter II, for proof]. The wavefront aberration W can be expressed as a power series in the pupil parameters x and y, the various terms of which can be

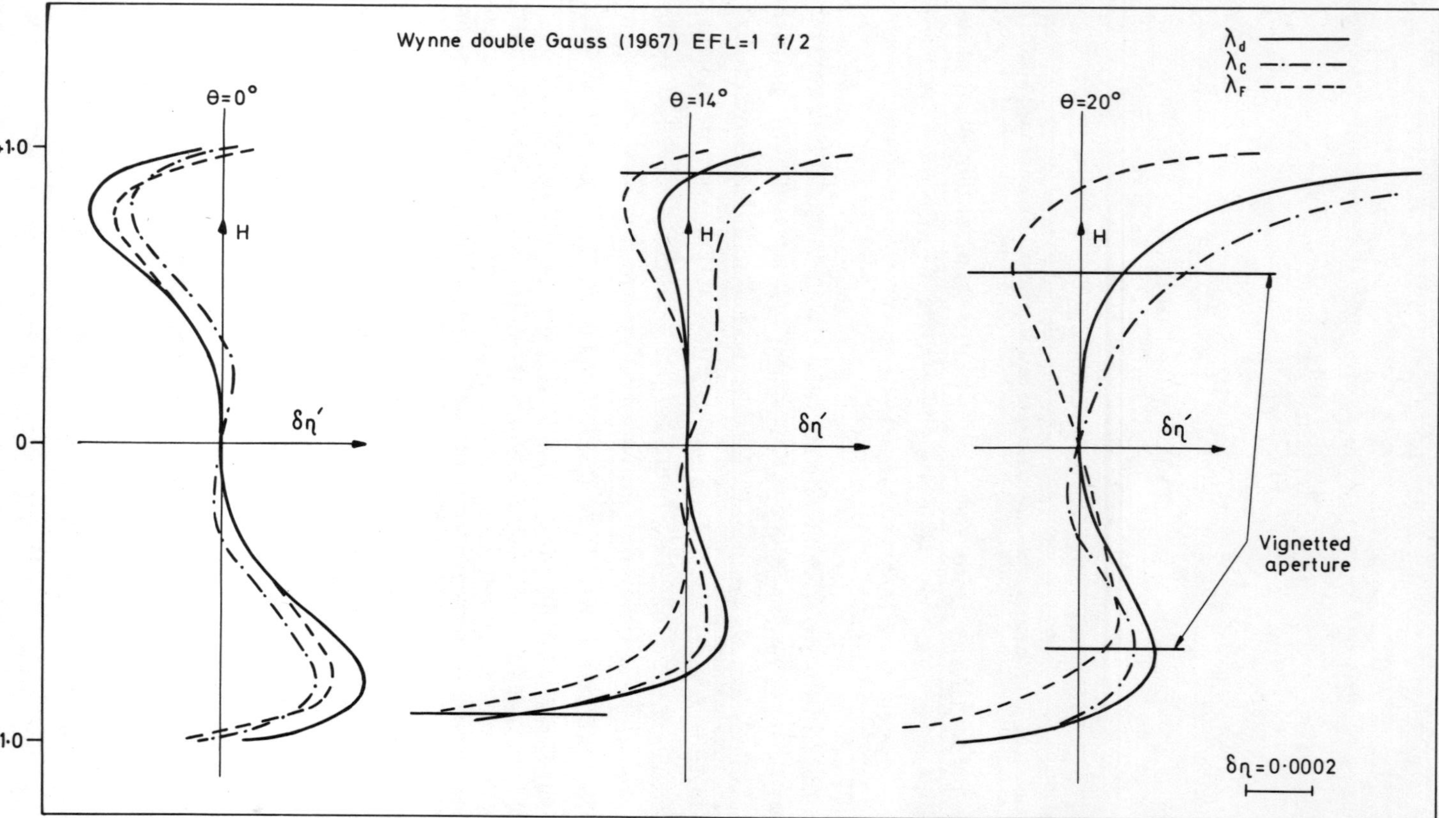

FIG. 11(*a*). Transverse ray aberration in the meridian section plotted for the Wynne double Gauss design.

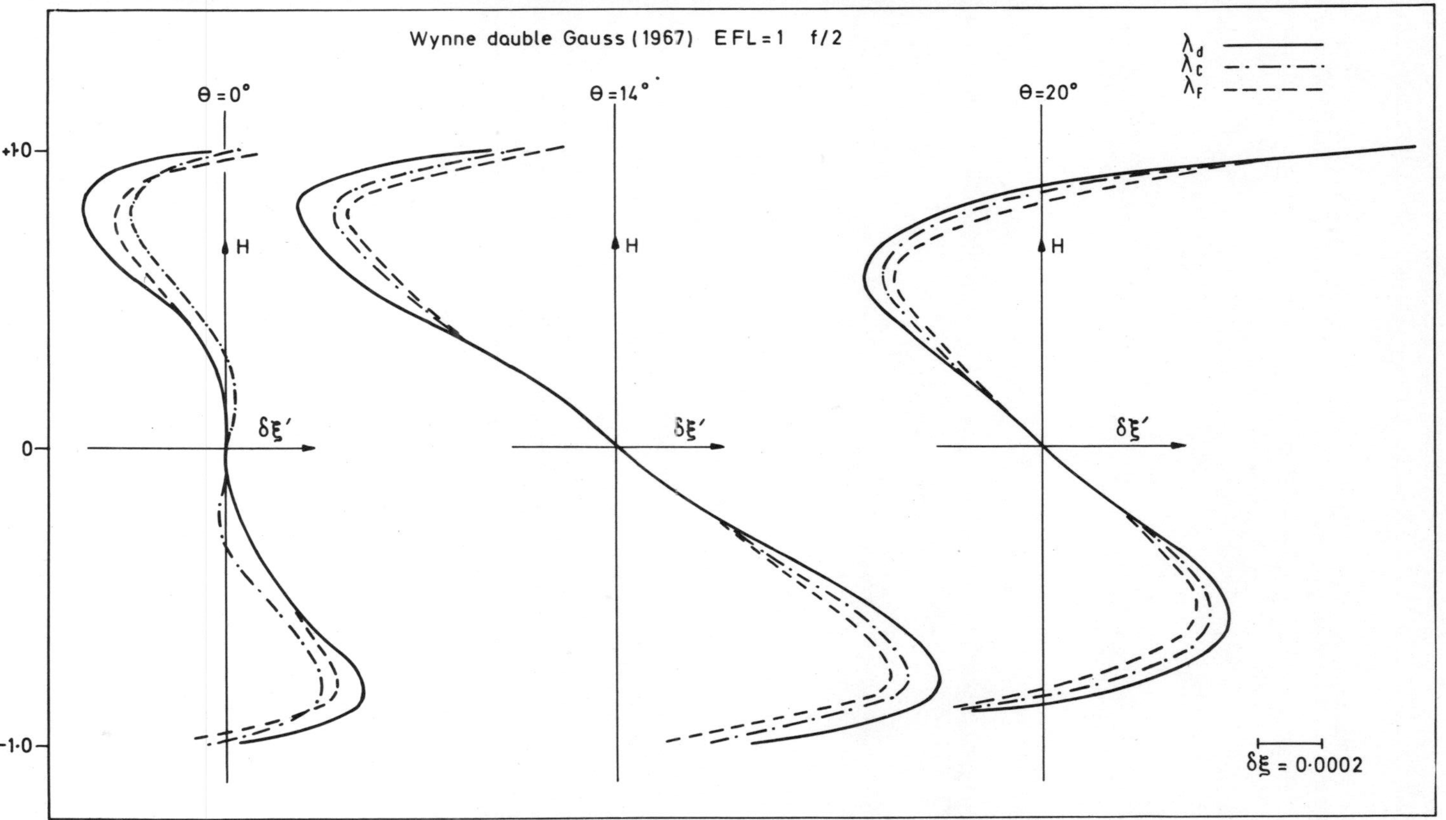

FIG. 11(*b*) Transverse ray aberration in the sagittal section—the $\delta\xi'$ component.

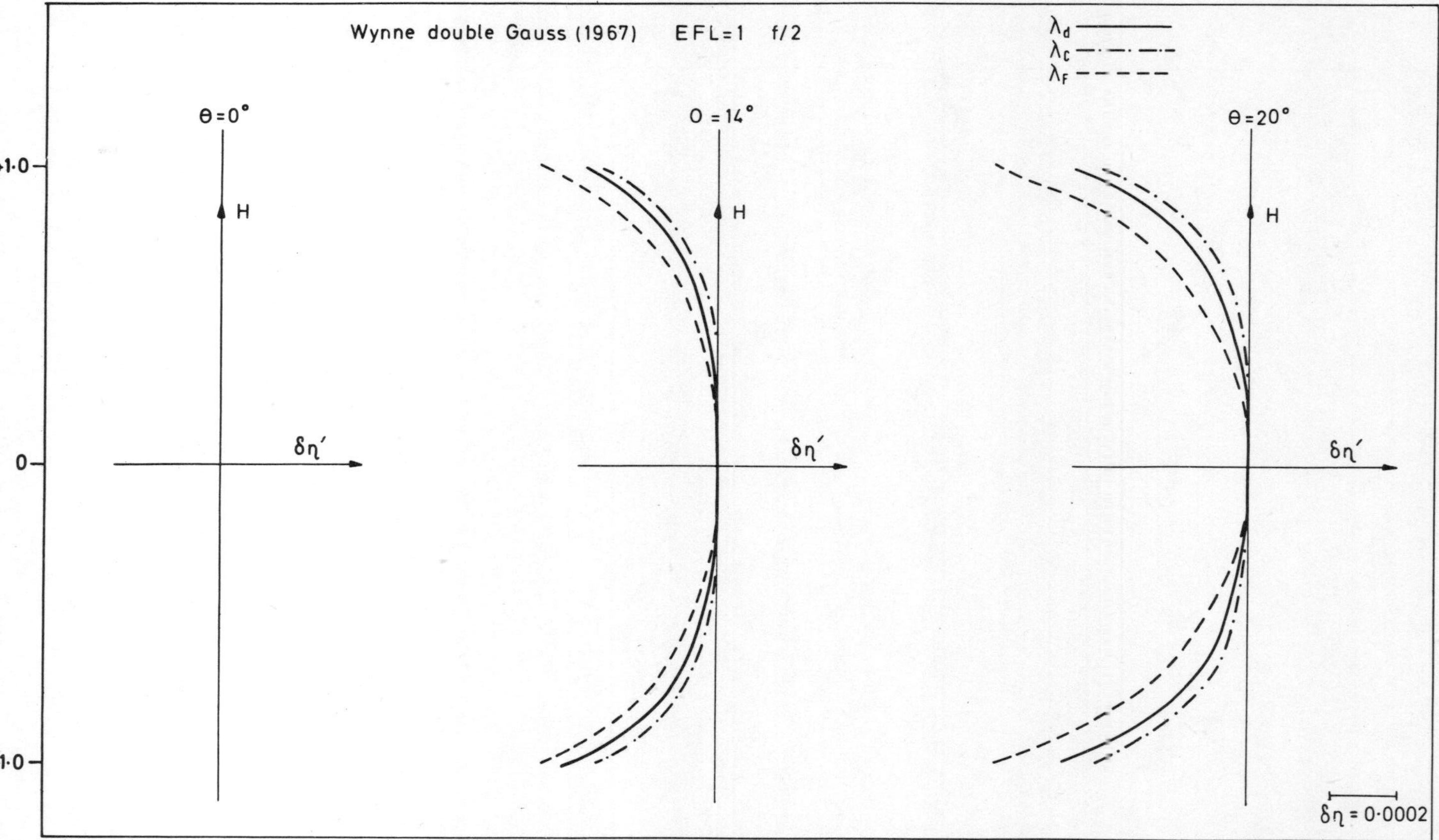

FIG. 11(*c*) Transverse ray aberration in the sagittal section—the $\delta\eta'$ component.

shown to represent the defocus, spherical aberration, coma and so on. It follows from this that defocus of the image plane corresponds to the addition of a linear term to transverse aberration or a rotation of the graph axis. It will be noticed that in Fig. 11 the axial aberrations for wavelengths other than the principal one have finite slopes at the origin resulting from the effect of longitudinal chromatic aberration. Similarly, primary spherical aberration will appear as a cubic term and primary coma as a parabolic term. Curvatures of the fourth power and higher represent the high order aberrations of the pupil and field. If we elect to plot the total transverse ray error, then, since distortion is a shift of the principal ray from the Gaussian image point, this error will manifest itself as a shift of the curve relative to the pupil ordinate. Similarly, chromatic difference of magnification or transverse chromatic aberration is indicated by the curves representing various wavelengths having different intersections with the aberration abscissa.

It follows that the designer either can treat transverse ray aberrations as a measure of the total image spread or by considering the shape of the aberration curves, as described above, he can determine the relative importance of the various aberration terms such as spherical aberration and coma. In the process of designing an objective the latter interpretation will be more satisfactory, whereas the first will determine whether the image spot size comes within the tolerance required by the specification of the optical system.

2.3.3. *Angular aberration*

Closely related to transverse aberration is angular aberration and for some purposes this form may be used in place of transverse aberration. It is sometimes employed in designing telescope objectives for use in measuring instruments (theodolites, spectrometers and astronomical telescopes) as it is the most direct measure of the instrumental error.

Suppose we consider the general case illustrated in Fig. 12. Let the point E′ be the centre of the exit pupil and the origin of the rectangular co-ordinates (X, Y, Z) and the point O′ be the

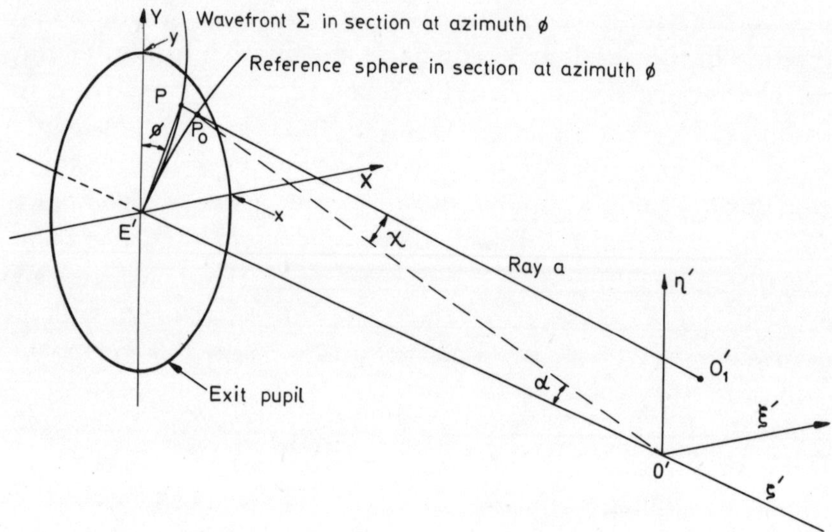

FIG. 12. Angular aberration of a skew ray in the image space.

axial Gaussian image point and the origin of the rectangular coordinate system (ξ', η', ζ'). We construct a sphere centred on O' with the radius of O'E' which we denote the reference sphere. Let Σ be the wavefront in the exit pupil E' (see § 2.5 for discussion of wavefronts). Then point P is the intersection of the skew ray *a* and the wavefront Σ. The line P_0O' is the normal to the reference sphere through the point P, that is, the direction of the ideal ray. The angular ray aberration is defined by the components of the angle between the ray a and the direction PO'. In order to analyse the extra-axial image points we replace the direction E'O' by the principal ray, thus tilting the diagram of Fig. 12 about the point E'. In Fig. 13 the ray b is in the meridian section and the angle χ is the angular aberration. Clearly, the angular aberration is approximately equal to the transverse aberration $\delta\eta'$ divided by the radius of the reference sphere R. Hence, it is not necessary to replot the meridian angular aberration for double Gauss objectives; we merely rescale the abscissa to read in angular measure. In Fig. 11(*a*) the transverse error of 0·001 is equivalent to 3 minutes of arc. We may also treat the angular errors of the sagittal section in a similar way by rescaling the plots shown in Figs. 11(*b*) and (*c*). Conrady (1929) considers the

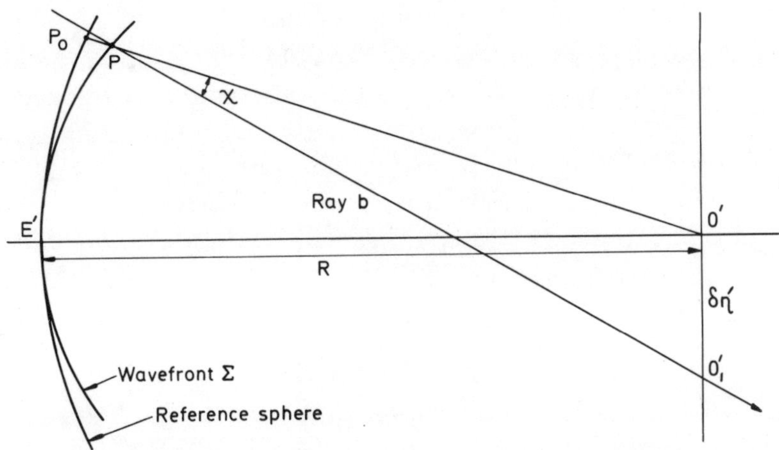

FIG. 13. Angular aberration of a meridian ray in the image space.

angular aberrations taken surface by surface as a measure of spherical aberration present in the objective. By making the appropriate trigonometrical approximations he obtained pure primary angular aberration, which could be calculated surface by surface with a form of paraxial ray tracing equations. This method does not seem to have achieved wide usage.

In some telescope systems with an infinite conjugate in the object space it is required to estimate the effect of the aberration in the object space from the image analysis of the image space. In this case the transverse aberration is divided by the effective focal length instead of the radius of the reference sphere.

2.3.4. *Aberration polynomials*

In § 2.2 we considered the generation of rays for use with electronic computers, a process by which the entrance pupil and the image plane are meshed by two grids of rays to chosen densities. When the available computer is sufficiently fast (an IBM 7094 traces five hundred rays surfaces per second), it is quite feasible to cover the image with a fine mesh consisting of several thousand rays. The mathematician, however, would regard this as a sledge hammer approach to the problem. There is an alternative approach.

First, finite rays of axial and several oblique pencils are traced and the transverse ray aberration or wavefront aberration calculated. Then it is possible to express the results in the form of an aberration polynomial, the variables of the polynomial being the pupil co-ordinates of the rays and the field angle of the principal rays of the oblique pencil. Consider, for example, a transverse ray aberration polynomial where we let the meridian and sagittal components of the transverse ray aberration be $\delta\eta'$ and $\delta\xi'$ respectively and the field angle or some other measure of the obliquity of the pencil be θ, and let x and y be the co-ordinates of the ray intersection in the pupil.

The functional relationships

$$\delta\xi' = P(x, y, \theta)$$
$$\delta\eta' = Q(x, y, \theta)$$

may be found by choosing polynomials of suitable type for P and Q and solving for the numerical coefficients by means of the ray tracing results. The number of terms to be taken in the polynomial depends on what orders of aberrations are expected to be significant; this depends on the kind of optical system and must be decided by trial and experience.

Conrady (1960) gives an example of an axial wavefront aberration polynomial, and Stravroudis and Feder (1954) give polynomials for transverse ray aberration, defining one polynomial per field angle. Herzberger (1947) also used aberration polynomials for generating spot diagrams (§ 2.3.5).

A graphic display can be generated from the aberration polynomial by using it to compute the ray intersection density and plotting this as a solid figure based on the image plans. If (x, y) are the pupil co-ordinates of a ray which meets the image plane at (ξ', η'), the ray intersection density on the image plane is proportional to

$$\frac{\partial(x, y)}{\partial(\xi', \eta')}$$

that is to say, this Jacobian represents the relative density of spots in a spot diagram or a geometrical point spread function.

F. Wandersleb (1952) and Stravroudis and Feder (1954) give diagrams representing these solids in two dimensional form and Welford (1963a) gives a perspective drawing for primary coma.

2.3.5. *Spot diagrams*

A spot diagram is a plot of ray intersection vectors in the image plane; each vector is represented by a spot at a distance from the origin proportional to the magnitude of the vector and its position according to the vector azimuth. In other words, the spot reconstructs the star image, neglecting the effect of the wave properties of light, at an enlarged scale and in such a way that the image intensity is proportional to the spot density. This form of aberration display was called 'spot diagram' by Herzberger (1947) but has frequently been used by Linfoot (e.g. Hawkins and Linfoot, 1945) in his publications over the last twenty-five years, but notably in his book *Recent Advances in Optics* (1955).

There are essentially two approaches to the generation of spot diagrams:

1. Using a fast electronic computer to trace a large number of rays (as described in § 2.2) and having the results plotted by means of an automatic plotting machine.
2. Using the technique of determining the transverse ray aberration polynomial (§ 2.3.4), from which any number of points of the spot diagram can quickly be obtained.

Hawkins and Linfoot (1945) used the first method whereas Herzberger (1947) and Stravroudis and Feder (1954) used the second.

In order that the spot diagram should be a faithful representation of the image point spread function, the fan of rays traced through the pupil for each field angle should represent a uniform flux of light. Therefore, it is important to ensure that the pupil is filled by a uniform density of rays by making a careful selection of rays in the ray generation procedure, taking into account any vignetting that may be present in the lens. The technique described in § 2.2.3 is suitable for this purpose. Other methods include:

1. A square grid superimposed on the entrance pupil, the ray intersection being the corners of the squares.
2. The recipolar array described by Cox (1964), which is the method based on concentric circles about the principal ray, the increment of the diameter of one circle to the next being equal for all circles.

3. An array of equilateral triangles used by Herzberger.
4. Concentric circles which are labelled in the image plane by programming the automatic graph plotter to provide a different 'spot symbol' for each ring of ray intersections in the entrance pupil.

Usually it is of some importance to evaluate the spot diagram in several focal planes about the paraxial focus in order to find the plane of best focus. This can be achieved without further ray tracing by the application of a simple defocus interpolation formula derived from similar triangles:

$$\frac{y' - \eta'}{L'} = \frac{y' - \eta'_t}{L' + t}$$

where L' is the distance from the exit pupil to the Gaussian image plane, t the longitudinal defocus displacement, y' the exit pupil co-ordinate, η' the ray height at the Gaussian image plane and η'_t the ray height at plane of defocus.

Then

$$\eta'_t = \eta' + \frac{(\eta' - y')t}{L'}$$

Let η'_p be the paraxial image of the object point and $\delta\eta'$ be such that

$$\eta' = \eta'_p + \delta\eta'$$

Similarly, the entrance pupil co-ordinate y is related to exit pupil co-ordinate y' by the relation $y' = my + \delta y'$, where m is the paraxial magnification of the image on the exit pupil plane of an object on the entrance pupil plane. Substituting,

$$\eta'_t = \eta' + \frac{t\eta'_p + t\delta\eta' - tmy - t\delta y'}{L'}$$

But $\delta\eta'$ and $\delta y'$ are small, and if t is small compared to L', then we can neglect $(t\,\delta y')/L'$ and $(t\,\delta\eta')/L'$. Since $t\eta'_p/L'$ is a constant representing a lateral distortion, it may also be neglected if the plot is not required to show distortion from the ideal image point.

Hence

$$\eta'_t = \eta' - \frac{tmy}{L'}$$

Similarly

$$\xi'_t = \xi' - \frac{tmx}{L'}$$

The interpretation of spot diagrams has been treated in detail by Herzberger (1957), who has developed, for large apertures and fields, a new type of image theory based, according to the increasing power of the aperture, on the grouping of aberration of the fifth degree. This approach leads to simple geometrical patterns which, he claims, enable the optical designer to ascertain the effects of changes in construction data on complex spot diagrams.

In making any interpretations of spot diagrams in terms of image intensity, it is important to remember that the effects due to the wave nature of light, that is to say, diffraction effects, are completely ignored. Herzberger (1947) has shown that for large aberrations, at least, these effects are negligible.

Nevertheless, the comparison of the spot spread diameter with the Airy disk is frequently made, and while this may be a useful guide to performance, it must be regarded with caution. In particular, the fine detail of the energy distribution is unlikely to follow the true diffraction image, except in the presence of gross aberration.

The spot diagrams of the double Gauss objective are shown in Plates 1-3.

2.3.6. *Kingslake transverse ray vector diagrams*

The interpretation of transverse ray errors for skew rays is not an easy matter. One approach, the spot diagram, as described in the previous section, has two disadvantages; namely, the difficulty of identifying the spot with its intersection point in the pupil, and the fact that the eye seems to weight displacement more heavily than the density of spots. For these reasons an alternative, though admittedly a more laborious, technique described by Kingslake (1966) seems to have some advantages.

Kingslake traced eighty-eight rays at a 12° obliquity through the system shown in Fig. 14 and plotted the ray displacements in the focal plane as a little vector for every ray (Fig. 15). The origin of each vector was taken to be the corresponding ray intersection in the pupil. The 22° obliquity is also shown in Fig. 15.

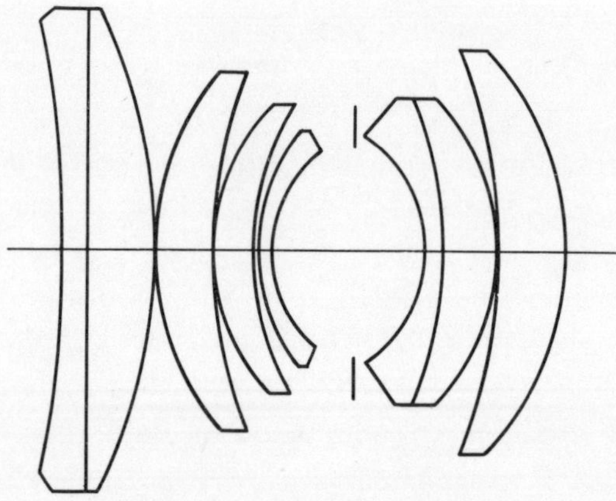

FIG. 14. An $f/1{\cdot}5$ double Gauss derivative due to Kingslake.

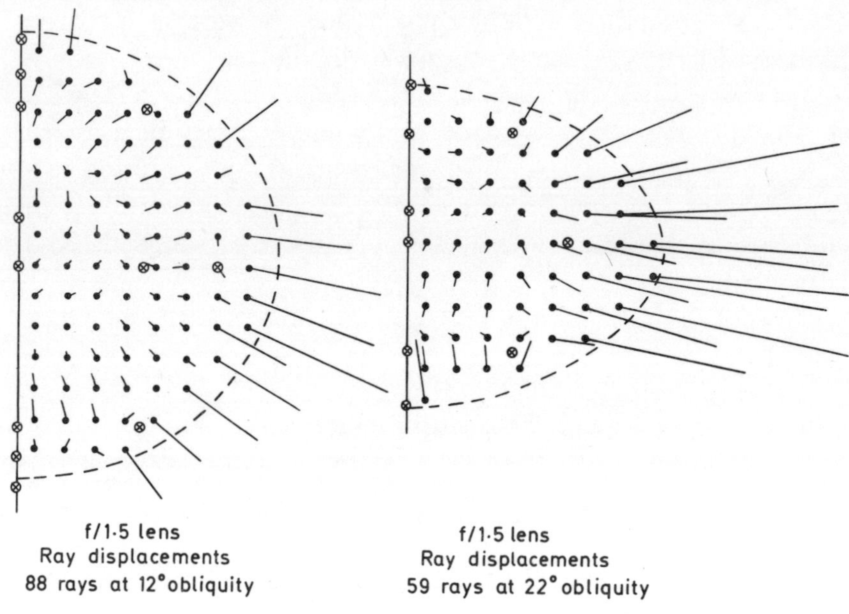

FIG. 15. Kingslake transverse ray vector diagrams.

Kingslake was concerned with the failure of a merit function to describe a system adequately (the rays defining the merit function are shown by a cross). Consequently, he was prepared to accept the labour of plotting involved, particularly as the method was not regarded as a routine display. It would not seem possible to automate the production of these diagrams as with spot diagrams, unless a comparatively elaborate computer program were devised. However, for the occasional extreme objective (such as the $f/1{\cdot}5$ double Gauss), the technique may prove useful.

2.3.7. *Diapoint aberration analysis*

Herzberger, who has contributed much to the study of spot diagrams, has for many years been an advocate of the use of diapoints to represent ray aberrations in the treatment of image formation by geometrical optics. The diapoint as defined by Herzberger (1936) is the point in the meridian plane formed by the intersection of a skew ray with the meridian section. This point with respect to some arbitrary terminal plane (such as the plane of best focus) has the co-ordinates $(0, y, z)$. This simple definition breaks down for meridian rays, because the rays that lie within the meridian plane cannot also form the diapoint intersection. In such an instance, the diapoint is defined as the sagittal focus, which is clearly the limiting position of the point as the ray transforms from a skew to a meridional ray.

If the optical system forms a perfect geometrical image, there is only a single diapoint. The presence of symmetrical aberration of the aperture (spherical aberration) causes the diapoints to lie along a line defined by the paraxial principal ray. In general, the diapoints will show asymmetry and so will be scattered about this line. The components of diapoint aberration can be written as $(\delta y, \delta z)$, corresponding to the displacement from some reference point $(0, y_0, z_0)$ in the terminal plane. The errors are plotted on a large scale giving δy as the ordinate against δz as the absisca, one plot per field angle.

We can see, therefore, that the diapoint bares the same relationship to longitudinal aberration as does the transverse aberration to the spot diagram. The co-ordinates of the diapoint $(0, y, z)$ are calculated from the ray trace co-ordinates $(x', y', 0)$ in the terminal plane for a ray having the direction cosines (L', M', N'),

$$y = y' - \frac{x'M'}{L'}$$

$$z = \frac{x'N'}{L'}$$

A special case arises where the image plane lies at infinity, as in the analysis of an ocular. Here the diapoints are found in a hyperthetical space, where the abscissa is plotted in units of reciprocal length of real space against the angular ray error. Herzberger explains this physically as equivalent to employing a perfect lens to form the diapoints in the vicinity of its focal plane.

The principal advantage of the diapoint representation over the transverse ray error is the independence of the plot from the choice of focal plane. A comprehensive study of diapoint analysis for photographic objectives has been undertaken by Hildebrand (1967). He obtained a new classification of aberration of ray errors based on diapoint imagery.

The relationship between diapoint errors and the normal classification in terms of the wavefront aberration coefficients is by no means immediately apparent from an examination of a typical set of errors. Broadly, however, we may say that the presence of spherical aberration leads to a spreading of the diapoints along a line which is either the z-axis or, for finite field angles, inclined to the z-axis at the field angle under investigation. Pure astigmatism also results in the diapoints being distributed along a line, but the diapoint error will contain a large y-component. Coma will cause the diapoint cluster to appear as a side view of a paraboloid dish.

When the origin of each ray in the pupil is known, useful information about the distribution of ray errors can be deduced if a line is constructed joining the diapoints obtained from a pupil scan parallel to the meridian section. A family of curves lying in the meridian section is thus obtained for a region around the chosen field point. These lines clearly demonstrate the vignetting of rays and the presence and origin of large ray errors which may require vignetting. The shape of the lines indicates the type of aberration terms present in the design.

2.4. IMAGE ANALYSIS—SPECIALIZED FORMS

The methods described in preceding sections are all concerned with the total image forming characteristics of rays traced through the system, but the optical designer is frequently interested in particular aberrations and so prefers a form of display which isolates these aberrations from the general image spread.

The total ray aberrations described in § 2.3 reduce to a special case in the formation of an axial image in a rotationally symmetrical system. This simplification arises because it is only possible to obtain aberrations of even powers of the pupil parameter such as defocus and spherical aberration. However, for extra-axial image points this symmetry is not available so the specialized forms described in the following sections are a valuable aid to the designer.

As an example, the designer of a profile-projection measuring instrument requires an objective with very low distortion. In the design process he will desire to obtain a clear idea of the distortion present so that he can apply the appropriate correction emphasis to this feature in trying to achieve an acceptable final balance of the aberrations.

2.4.1. *Offence against sine condition*

The sine condition was discovered almost simultaneously but quite independently in 1873 by Abbe and Helmholtz, with slight priority of publication for Abbe. (An earlier sine theorem in thermodynamics was not recognized in optics until some decades later.) In its original form it proved useful, up to the time of Conrady, in establishing aplanatism, in the Abbe sense of the term, for rotationally symmetrical systems.

At that time ray trace calculations were executed at the expense of some considerable labour using six-figure logarithms, so that the designer arranged his calculations in order that he could extract the greatest possible amount of information from every ray traced. In this respect, Conrady was particularly notable in his contribution to the art of calculation, as can be seen in his great work on optical design (Conrady 1929, 1960). It was Conrady who developed a considerably more useful form of the optical sine theorem which he called offence against sine condition (OSC′).

Here again the effortless and fast ray tracing possible with electronic computers has caused a change to take place in optical design methods and to some extent lessening the value of techniques such as OSC′. Nevertheless, it is a particularly elegant device and since it also possesses some interesting properties connected with coma the following discussion is included.

In Fig. 16 the object OO_1, of height η, is imaged paraxially at $O'O_1'$ by the system with entrance and exit pupils E and E′

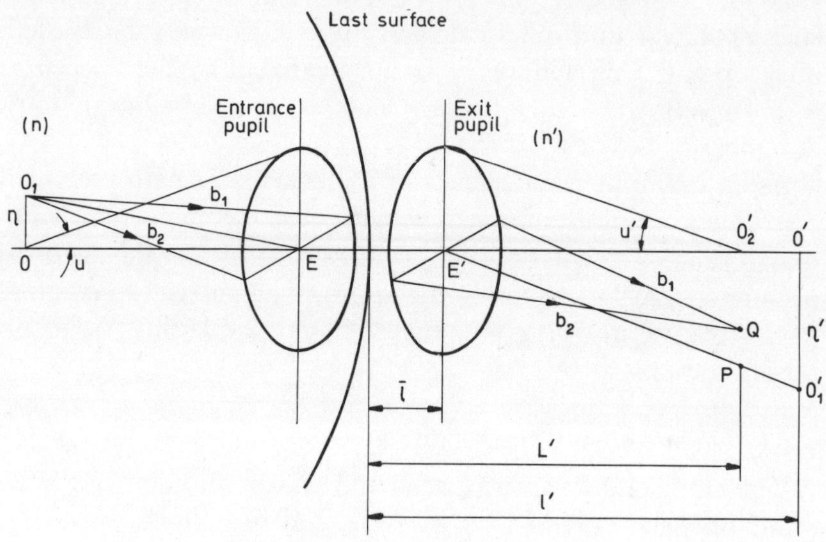

FIG. 16. Sagittal coma.

respectively. Let these pupils subtend the paraxial convergence angles u and u'. A sagittal fan of rays b_1b_2 of finite aperture is imagined at Q. The perpendicular to the optical axis at O_2' passes through Q and P, the intersection with the principal ray. Let a finite ray from O, traced in the meridian section, have convergence angles U and U' in the object and image spaces, respectively. In terms of ray tracing equations

$$\sin U = -M$$

$$\sin U' = -M'$$

Conrady defined sagittal coma as the distance QP and to obtain a measure independent of linear scale took the ratio $QP/O_2'P$ for the OSC′. Hence, with the notation of the diagram he showed that

$$\text{OSC}' = 1{\cdot}0 - \frac{\sin U}{u} \cdot \frac{u'}{\sin U'} \cdot \frac{l' - \bar{l}'}{L' - \bar{l}'} \tag{2.1}$$

OSC′ is therefore a measure of transverse ray coma. In the absence of spherical aberration the term $(l' - \bar{l}')/(L' - \bar{l}') = 1{\cdot}0$ and when $\text{OSC}' = 0$, we obtain

Plates 1–3

Spot diagrams for a double Gauss objective (see p. 31).

Plates 4–6

Interferograms for a Wynne double Gauss system (see p. 105).

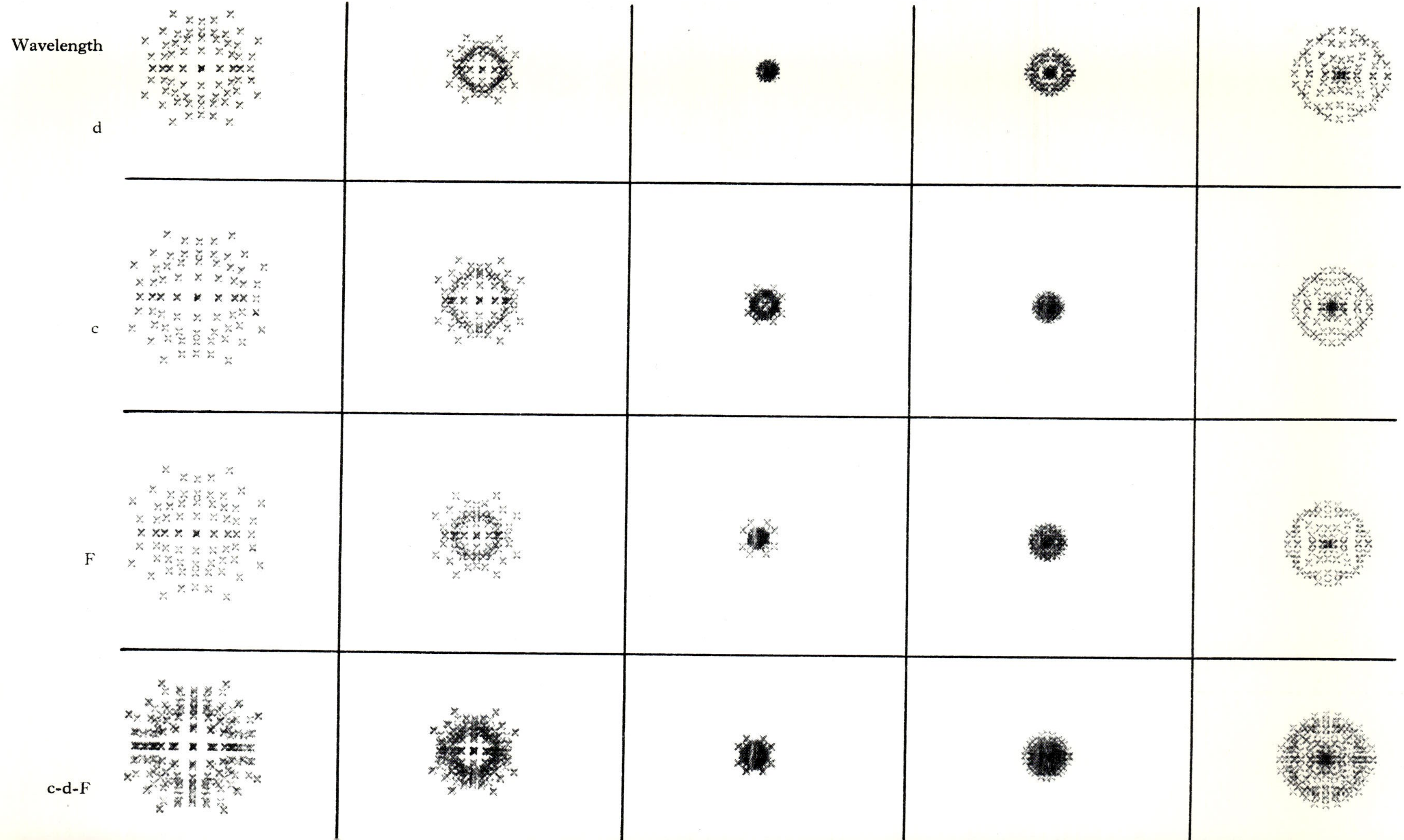

Wavelength

d

c

F

c-d-F

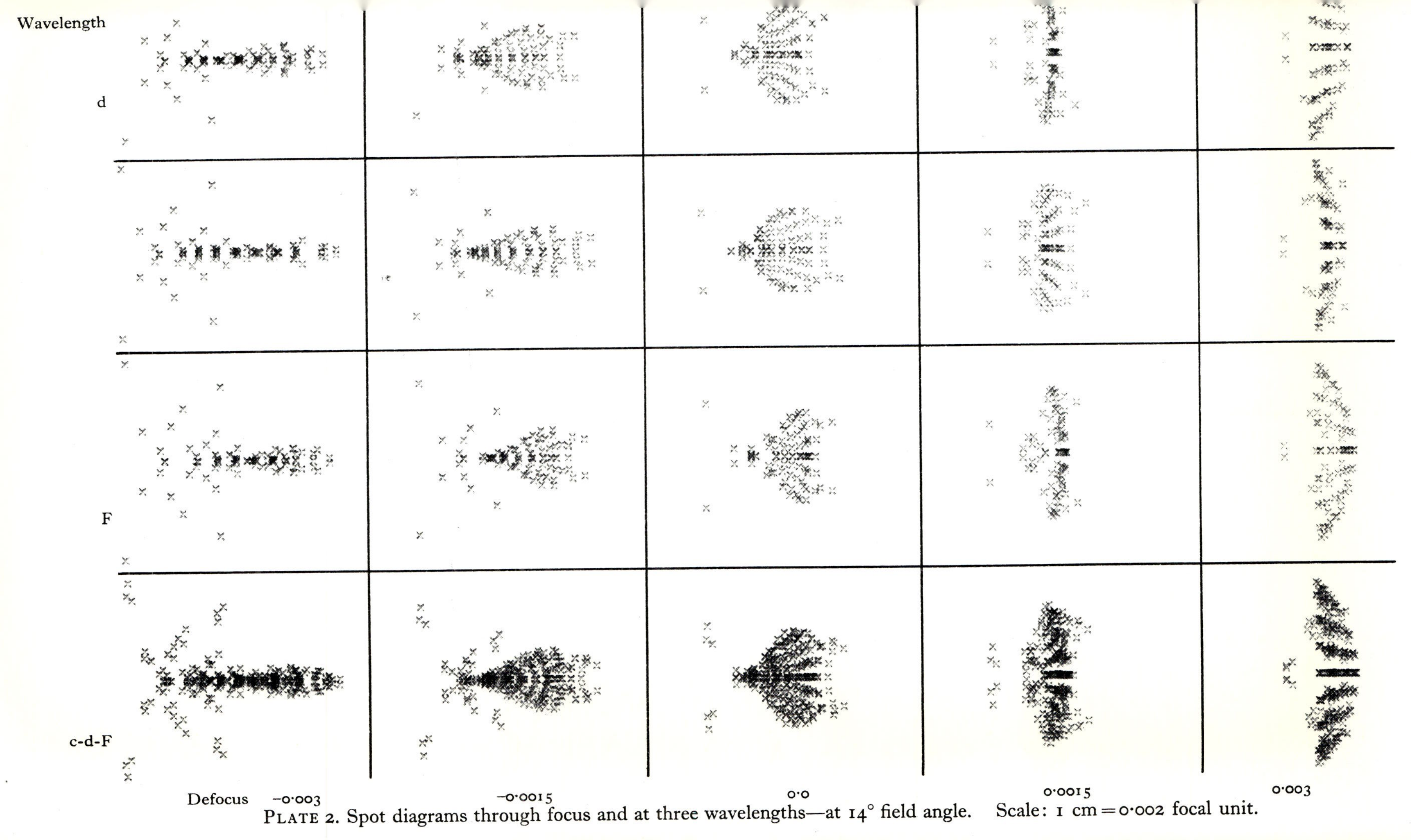

PLATE 2. Spot diagrams through focus and at three wavelengths—at 14° field angle. Scale: 1 cm = 0·002 focal unit.

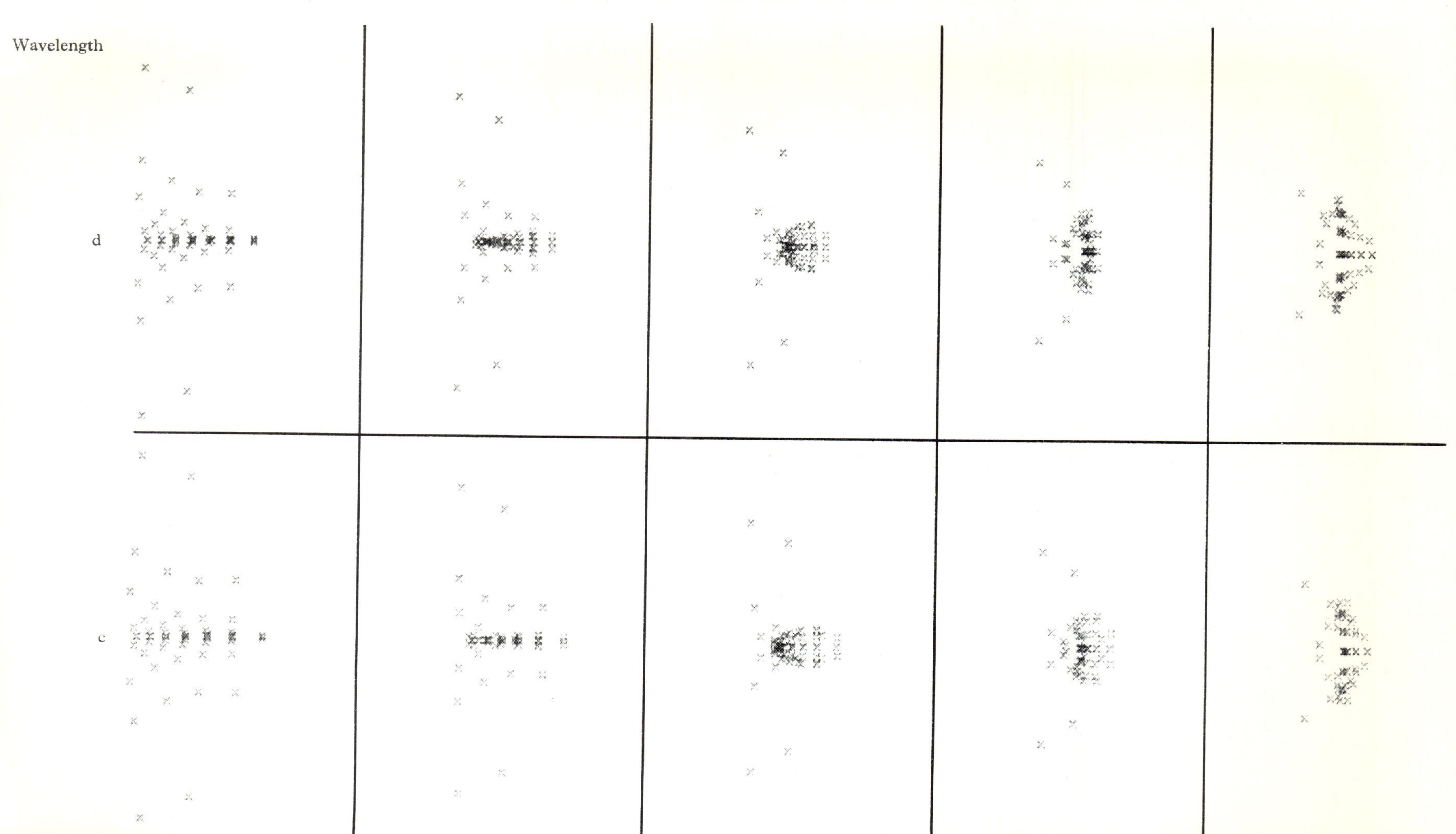
Wavelength
d
c

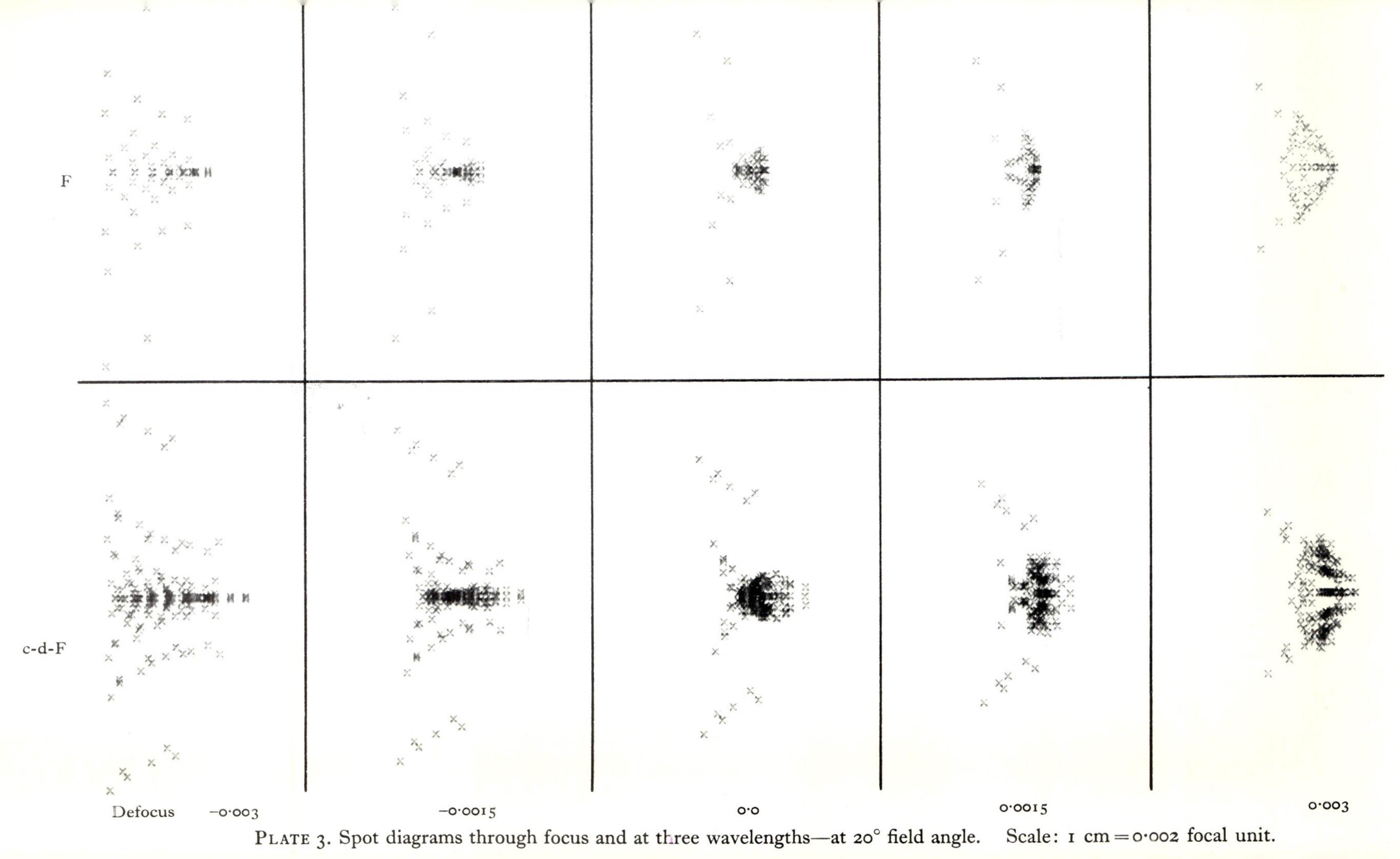

PLATE 3. Spot diagrams through focus and at three wavelengths—at 20° field angle. Scale: 1 cm = 0·002 focal unit.

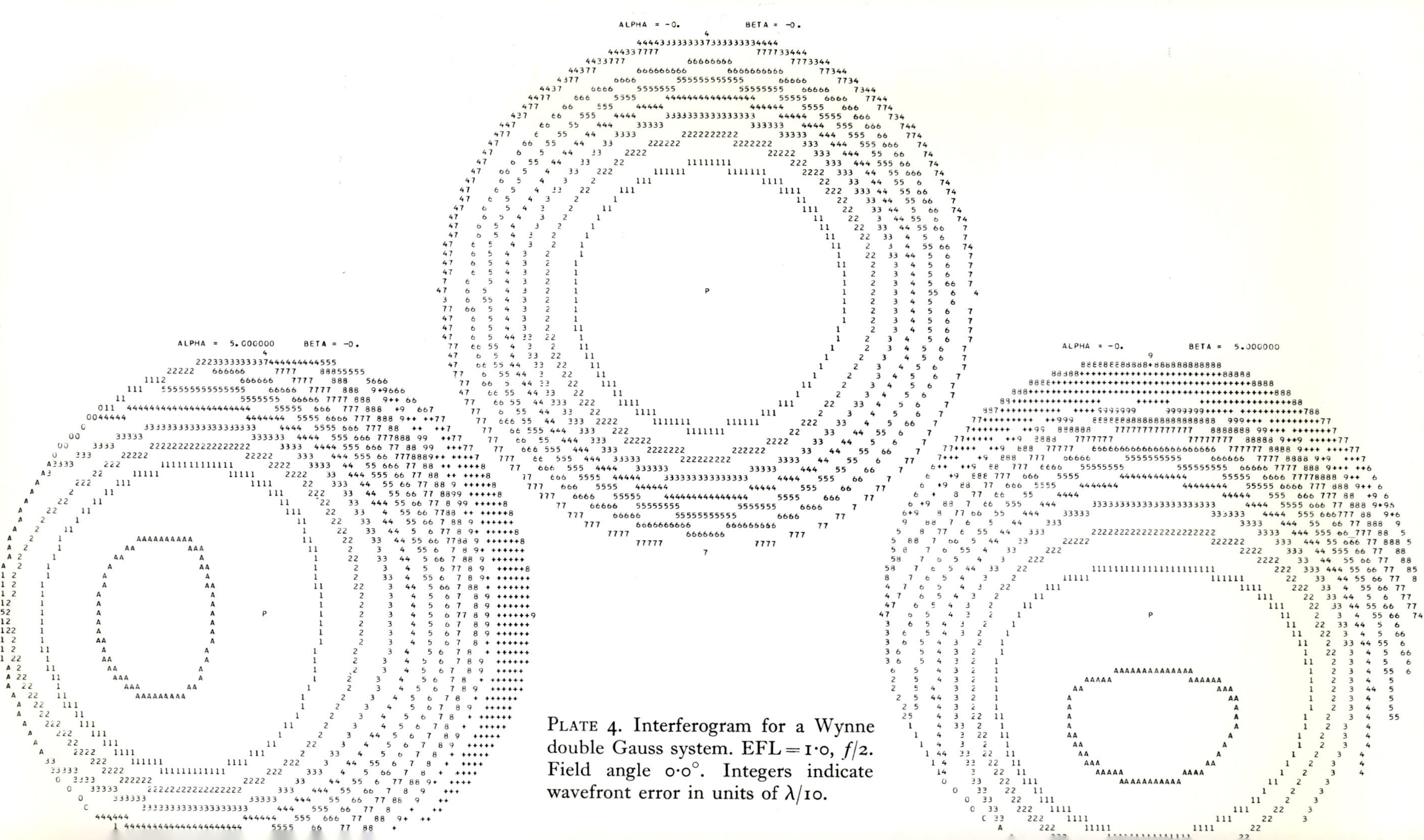

PLATE 4. Interferogram for a Wynne double Gauss system. EFL = 1·0, $f/2$. Field angle 0·0°. Integers indicate wavefront error in units of $\lambda/10$.

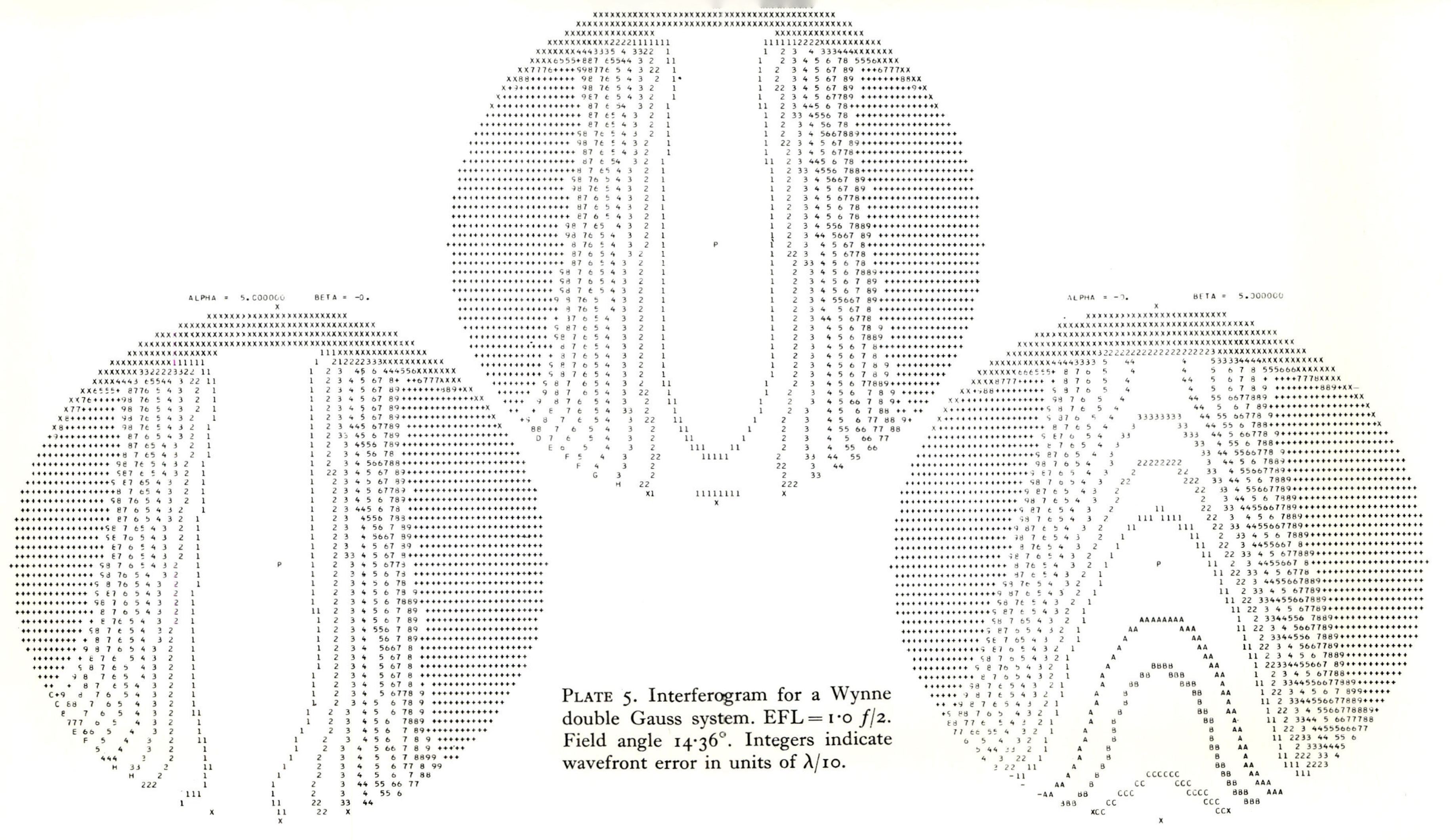

PLATE 5. Interferogram for a Wynne double Gauss system. EFL = 1·0 $f/2$. Field angle 14·36°. Integers indicate wavefront error in units of $\lambda/10$.

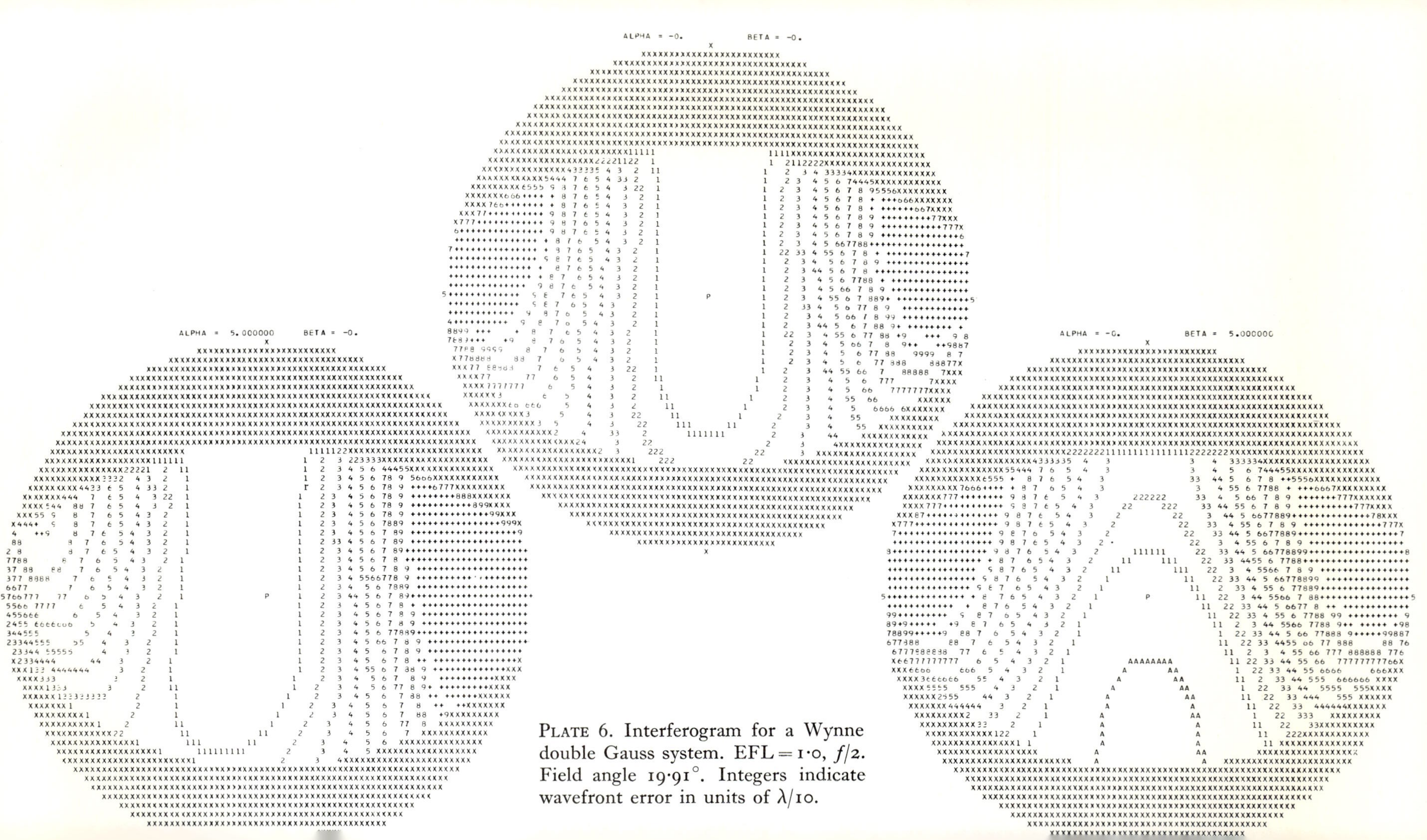

PLATE 6. Interferogram for a Wynne double Gauss system. EFL = 1·0, $f/2$. Field angle 19·91°. Integers indicate wavefront error in units of $\lambda/10$.

$$\frac{\sin U}{\sin U'} = \frac{u}{u'} \quad \text{the Abbe sine condition}$$

For objects at infinity the expression $\sin U/u$ becomes indeterminate. Replacing it by Y/y, where Y is the finite ray height at the first surface and y the paraxial marginal ray height at the first surface, permits the calculation of OSC′ for this special case.

Some authors (for example, Cox, 1964) adopt a simplified form which assumes zero spherical aberration when $L' - \bar{l}' = l' - \bar{l}'$; the equation therefore becomes:

$$\text{OSC}' = 1{\cdot}0 - \frac{\sin U}{u} \; \frac{u'}{\sin U'}$$

Actually, the expression as given by Cox is not derived from the Conrady definition of OSC′ but from a comparison of the transverse magnifications of a meridian section finite ray and a paraxial ray. So Cox's definition of OSC′ treats coma as a difference of traverse magnification of the finite rays in various zones of the meridian section.

From the Lagrange relation

$$nu\eta = n'u'\eta'$$

Hence the paraxial magnification

$$m' = \frac{\eta'}{\eta} = \frac{nu}{n'u'}$$

and the finite ray magnification

$$\text{Mag} = \frac{n \sin U}{n' \sin U'}$$

The OSC′ is then given as

$$\text{OSC}' = \text{Mag} - m'$$

For a finite object conjugate

$$\text{OSC}' = \frac{n \sin U'}{n' \sin U'} - \frac{nu}{n'u'}$$

and for an infinite object conjugate

$$\text{OSC}' = \frac{n\,Y}{n' \sin U'} - f' \qquad (2.2)$$

where f' is the effective focal length in the image space.

Yet another form, variously termed and which we shall call sine condition coma (W'_{coma}), derives from the wavefront theory of aberration. Hopkins (1946 and 1952) showed that it is possible to write the coefficient of a wavefront coma term as

$$W'_{\text{coma}} = H\left(\frac{\sin \theta'}{u'} - \frac{\sin \theta}{u}\right)$$

where θ and θ' are angles of aperture which in the absence of spherical aberration reduce to $\theta = U$ and $\theta' = U'$, again giving the sine condition when $W'_{\text{coma}} = 0$.

Substituting for parameters obtained from ray tracing we obtain

$$W'_{\text{coma}} = H\left[\frac{\sin U'}{u'}\left(1{\cdot}0 - \frac{\delta L'}{l' - \bar{l}'} \cos U'\right) - \frac{\sin U}{u}\right]$$

We note that it is necessary to put

$$\frac{\sin U}{u} \rightarrow \frac{Y}{y}$$

when $U = 0$. Also putting $\cos U' \rightarrow 1{\cdot}0$ for small U' and $\delta L' = L' - l'$ we obtain

$$W'_{\text{coma}} = H\left[\frac{\sin U'}{u'}\left(\frac{L' - \bar{l}'}{l' - '\bar{l}}\right) - \frac{\sin U}{u}\right]$$

We can easily relate this to the Conrady OSC′ by removing a term from the parenthesis:

$$W'_{\text{coma}} = H\left(\frac{L' - \bar{l}}{l' - \bar{l}'}\right)\frac{\sin U'}{u'}\left(1{\cdot}0 - \frac{\sin U}{u} \; \frac{u'}{\sin U'} \; \frac{l' - l'}{L' - \bar{l}'}\right)$$

$$W'_{\text{coma}} = H\left(\frac{L' - \bar{l}'}{l' - \bar{l}'}\right)\frac{\sin U'}{u'} \text{ OSC}'$$

The term in parenthesis represents the spherical aberration of the pupil.

Wavefront aberration is treated more fully in § 2.5, but if here we consider the wavefront coma terms it will help us to see the usefulness of OSC′. Coma and high order coma-like terms of the pupil can be expressed in a general form as the sum

$$W'_{\text{coma}} = \sum_{n=1}^{\infty} C_n \, \rho^{2n+1} \, \eta \cos \phi$$

where C_n is a coefficient, ρ the pupil height, η the field height, and ϕ the azimuth.

In the derivation of the expressions for OSC′ and W'_{coma} (Conrady and Hopkins), the important assumption was that the field height η was sufficiently small to allow terms of higher order than the first in η to be neglected. On the other hand, the finite aperture angles are included; that is, we only consider higher order coma terms in the pupil but not those in the field. For this reason OSC′ gives a considerably more precise measure of coma than Siedel S_{II} sums and yet requires little extra calculation beyond that involved in tracing the meridian ray. Presented graphically, the normalized pupil height H is plotted as ordinate against OSC′ as abscissa. We note that from the symmetry about the optical axis we need only plot the semi-aperture. Fig. 17 shows two forms of plot for the double Gauss example, the first obtained by using equation (2.1) and the second by using the approximate form of (2.2).

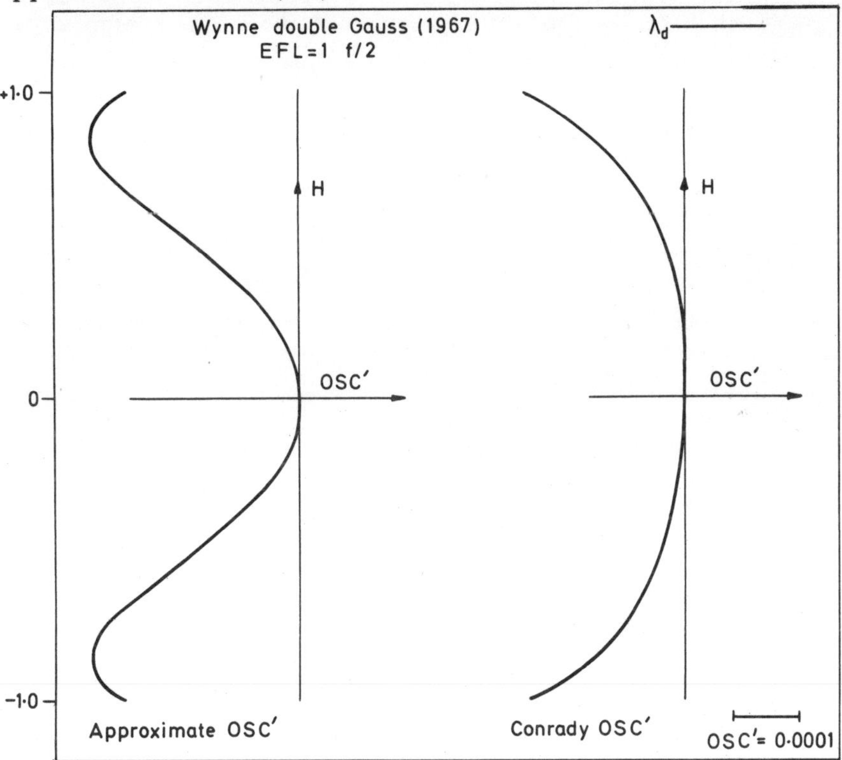

FIG. 17. The Conrady OSC′ plotted for the Wynne double Gauss design.

2.4.2. *Sagittal and tangential ray tracing*

Astigmatism in a system is usually examined by the technique of sagittal (s) and tangential (t) ray tracing. As with OSC′ discussed in the previous section, this technique arose in the days of manual ray tracing methods, with economy of labour in mind.

A principal ray is traced using the meridian form of the recurrence relationships discussed in § 2.1, and then the additional calculations required to trace the s and t rays amount to little more than that of a paraxial ray trace. In other words, the s and t rays correspond to rays traced about the principal ray in the sagittal and meridian sections respectively, to the Gaussian approximation which implies the recurrence relationships involve only the first order in the pupil parameter ρ. We note that this technique contrasts with OSC′ which derives from consideration of the sum of coma terms of all orders of the pupil parameter but only the first order in the field parameter η, whereas in tracing s and t rays we consider vanishingly small apertures in the pupil but finite field heights. It follows that we neglect high order pupil astigmatisms. In spite of this limitation the technique is still in common use and is considered to be a valuable guide to the astigmatism of a system.

Hopkins (1946, 1952) describes a set of recurrence relationships (actually transformations of an earlier equation given, for example, by T. Young and von Rohr) which allow the determination of the tangential and sagittal focus for each principal ray. The distance $\bar{D}$ between surfaces measured along the principal ray is calculated from the data of a meridian ray trace using the equation

$$\bar{D}=\frac{1}{N}(d-z_{-1}+z)$$

Sagittal fans are traced using the relation

$$\text{Transfer } h_s=h_{s-1}-\bar{D}u_s$$

$$\text{Refraction } u'_s=\frac{1}{n'}[nu_s+h_s c(n'\cos\bar{I}'-n\cos\bar{I})]$$

Tangential fans are traced using the relations

$$\text{Transfer } h_t=\frac{h_{t-1}\cos\bar{I}'_{-1}-\bar{D}u_t}{\cos\bar{I}}$$

$$\text{Refraction } u'_t=\frac{1}{n'\cos\bar{I}'}[n\cos\bar{I}u_t+h_t c_t(n'\cos\bar{I}'-n\cos\bar{I})]$$

where h_s and h_t are the ray heights measured from the principal ray and c_t is the curvature of surface at the point of interception with the ray in the tangential section (necessary for aspheric surfaces).

The initial ray data are taken from principal ray data at the first surface $h_s = h_t = y$

$$-u_s = \frac{M}{(1 - M^2)^{\frac{1}{2}}}$$

$$-u_t = \frac{M}{(1 - M^2)^{\frac{1}{2}}}$$

The final transfer is used to calculate the sagittal and tangential focal errors s and t measured along the optical axis:

$$s = \left(\frac{h_s}{u_s} - \bar{D}\right) N$$

$$t = \left(\frac{h_t \cos I'}{u_t} - \bar{D}\right) N$$

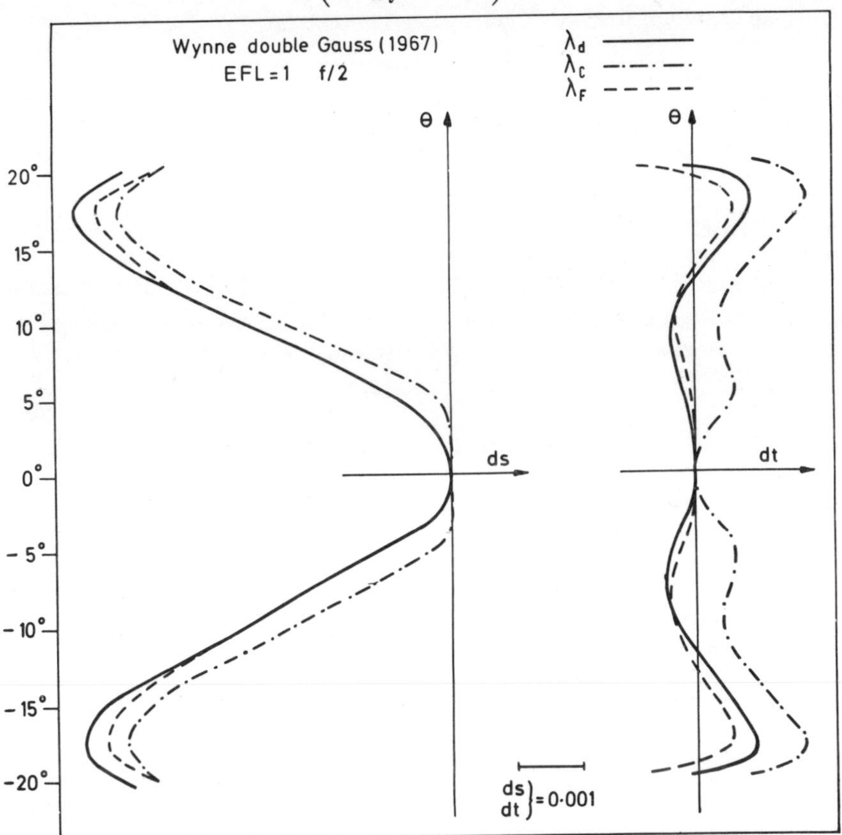

FIG. 18. Astigmatism demonstrated by sagittal and tangential ray errors for the double Gauss design.

Graphically, the sagittal and tangential focal errors are usually plotted as abscissa against the field angle as ordinate, as in Fig. 18.

An alternative method of using the *s* and *t* data, particularly informative when used with the plot of transverse ray aberration, is plotting the *s*-slope and *t*-slope. This form represents the slopes of the transverse ray aberration curve at the origin of the graph, i.e. the point corresponding to the principal ray intersection with the Gaussian image plane. The *s*-slope is the gradient of the aberration curve in the sagittal section and the *t*-slope that of the meridian (tangential) section. Clearly for this purpose we require the transverse aberration of a ray traced at a limitingly small aperture about the principal ray in the appropriate section, and this is precisely what the *s*- and *t*-traces can supply. The final transfer in this case is used to calculate the transverse error:

The transverse error in the sagittal section

$$ds = \left(\frac{h_s}{u_s} - \bar{D}\right) u'_k$$

The transverse error in the tangential section

$$dt = \left(\frac{h_t \cos \bar{I}}{u_t} - \bar{D}\right) u'_k$$

where u'_k is the convergence angle of the image ray at the image plane. The values of ds and dt are plotted as transverse aberrations at the full pupil height, where $H = 1$ corresponding to pupij aperture of $\tan (u'_k)$. The straight lines joining the points $(dt, \tan u'_k)$ and $(ds, \tan u'_k)$ to the origin represent the slopes of the transverse aberration. If the transverse aberration has been calculated relative to the Gaussian image point, then the displacement $\delta\eta'$ of the principal ray should be added to the transverse tangential error and the point $(dt + d\eta', \tan u')$ joined to the point representing the principal ray $(\delta\eta', 0)$.

In the absence of astigmatism of the first order in the pupil and all orders in the field the two slopes are equal, i.e. the plotted *s*-slope coincides with the *t*-slope. In the absence of field curvature they also coincide with the pupil ordinate.

The *s*- and *t*-slopes are of great assistance in drawing accurately the transverse aberration curves, but also give useful information about astigmatism and field curvature.

2.4.3. *Percentage distortion*

Image distortion is defined as the distance between the point of intersection of the principal ray with the meridian ray and the image point defined by Gaussian imagery. Let this distance be $\delta\eta$ and the height of the Gaussian image be η', then the percentage distortion is defined by $(\delta\eta'/\eta') \times 100$.

We have seen in § 2.3.2 that the transverse ray aberration can be defined to include or omit the distortion term depending on whether the error is measured with respect to the Gaussian image point or to the principal ray intersection. In the latter case, particularly, it is necessary to display the distortion as a separate plot, but in either case it is useful to be able to isolate the distortion of the image.

Unfortunately, in a practical system, it is exceedingly difficult to measure distortion in terms of the above definition as there is no way of identifying the principal ray in the image plane. Also, as a result of non-linearities of the observed image such as caused by the presence of coma, this definition is not necessarily a realistic indication of the centre of the intensity distribution of any point image. In a more realistic definition it would be necessary to replace the principal ray intersection with the coordinates of the point of greatest intensity as given, by say, a spot diagram. Of course, this cannot be used in general as it is much more troublesome to calculate, so the original definition is used as a good approximation. The percentage distortion for the double Gauss example is shown in Fig. 19.

2.5. GEOMETRICAL WAVEFRONT ABERRATION

Suppose that a number of rays are traced from a point in the object space through the system to the image space. By considering the optical pathlengths along each ray, and applying Fermat's principle, we can construct a series of surfaces in the image space over which the phase (optical path) is constant. Any one of these surfaces is called a wavefront, and the asphericities of the image forming wavefronts are the significant quantity in determining the quality of an image. This gives rise to the concept of wavefront aberration. In Fig. 12 (p. 26) let the point E′ be the centre of the exit pupil and the origin of the rectangular co-ordinate system (X, Y, Z), and let the point O′ be the axial Gaussian image point

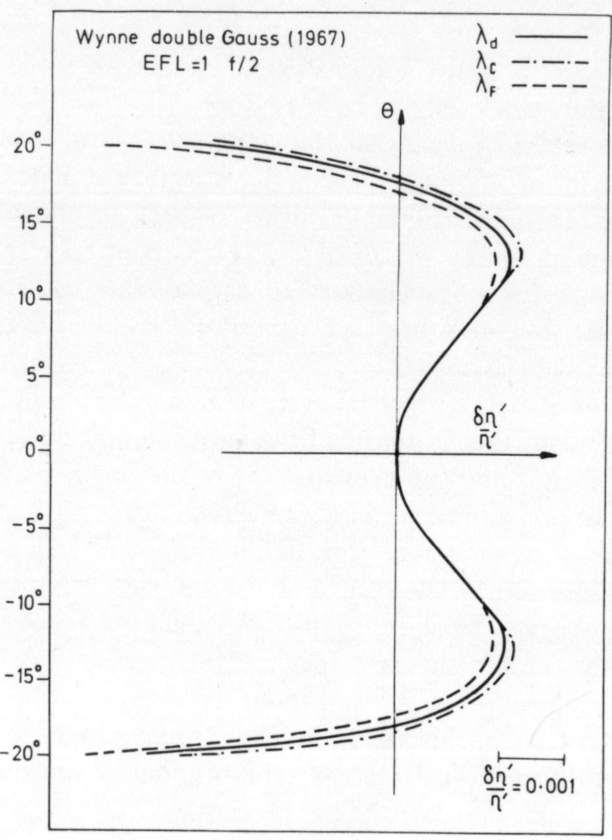

FIG. 19. Principal ray distortion plotted for Wynne double Gauss design.

and the origin of the rectangular co-ordinate system (ξ', η', ζ'). We construct a sphere centred on O′ with the radius O′E′ which we denote 'the reference sphere'. Let Σ be the wavefront in the exit pupil E′, and let the ray a be a typical skew ray which will be a normal to the wavefront. The point P is the intersection of the skew ray a and the wavefront Σ, and the line P_0O' is the normal to the reference sphere through the point P. The wavefront aberration is equal to the optical path PP_0. For extra-axial image points the reference sphere is centred on the point of intersection of the principal ray and the Gaussian image plane.

2.5.1. *Calculation of wavefront aberration*

Conrady (1960) treats the calculation of the wavefront aberration for an axial pencil, and Hopkins (1950) for rays in the meridian section. In a later paper, Hopkins (1952) gives a method for calculating the wavefront aberration of skew rays.

Skew rays present a difficulty, for it is quite possible that the principal ray and skew ray will never intercept to produce a focus, that is, a point for which the aberration remains constant as the wavefront progresses. Hopkins' method was to define invariant foci for which the aberration does remain constant with wavefront propagation, and to show that the mid-point of the shortest join between two skew rays was an invariant focus. (There exists, in fact, an infinity of invariant foci lying in a straight line.) The wavefront aberration for each ray was then referred from the invariant focus to the centre of the reference sphere.

The obvious method for calculating the optical path along any ray would be to sum the values of the expression $(n \, . \, D)$ between successive surfaces from the object to the focus, the values of D being obtained from a finite ray trace (§ 2.4.2). Similarly, the path along the principal ray $\Sigma n\bar{D}$ could be found and hence the wavefront aberration

$$W = \Sigma nD - \Sigma n\bar{D}. \qquad (2.3)$$

Unfortunately, this procedure of differencing two large quantities leads to poor numerical accuracy, and an alternative must be sought. Hopkins' method neatly overcomes the problem by calculating the aberration from the co-ordinates and the direction cosines produced by the skew and principal rays at each surface as the ray trace proceeds.

The method is summarised here:

Let the aberration produced by an interface between two isotropic media be $dW = W' - W$, where W' and W are the aberrations along the given ray, relative to the principal ray of the incident and emergent surfaces. In Fig. 20, let $\bar{\mathrm{Q}}\mathrm{Q}$ and $\bar{\mathrm{Q}}'\mathrm{Q}'$ be the two reference spheres with centres F and F′, the mid-points of the shortest join of the incident and emergent rays respectively. The aberration due to the interface is given by

$$\delta W = [\bar{\mathrm{Q}}\bar{\mathrm{P}}\bar{\mathrm{Q}}'] - [\mathrm{QPQ}'] \qquad (2.4)$$

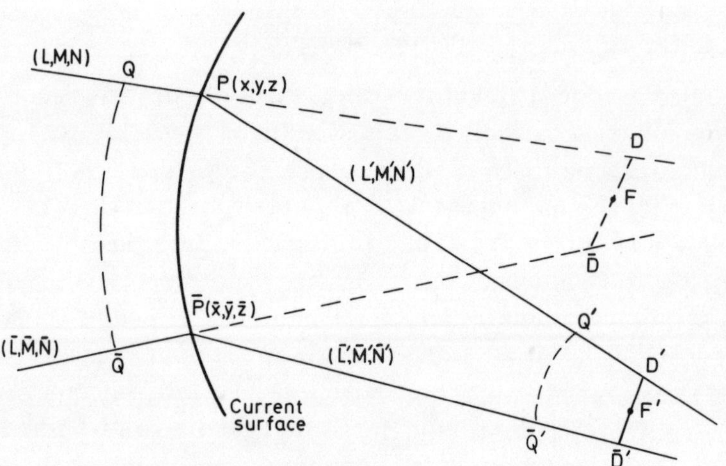

FIG. 20. The propagation of wavefront aberration according to Hopkins' conventions.

The square brackets used here indicate optical pathlength, and P̄ and P are the points of incidence of the two rays at an interface separating media of refractive index n and n'. Q̄′Q and Q̄′Q′ are great circles of spheres with centres F, F′ and consequently,

$$QD=\bar{Q}\bar{D},\; Q'\bar{D}'=\bar{Q}'D'.$$

Expression (2.4) is therefore equivalent to

$$\delta W=\Delta(ne)$$

where $e=\bar{p}-p$, and $\bar{p}=\bar{P}\bar{D}$ and $p=PD$. Let $(\bar{L}, \bar{M}, \bar{N})$ and (L, M, N) be the direction cosines of the two incident rays, and let $(\bar{X}, \bar{Y}, \bar{Z})$, (X, Y, Z) be the rectangular co-ordinates of the points of incidence P̄, P. Then:

$$\bar{p}=\bar{L}(\bar{x}-\bar{X})+\bar{M}(\bar{y}-\bar{Y})+\bar{N}(\bar{z}-\bar{Z})$$
$$p=L(x-X)+M(y-Y)+N(z-Z)$$

where $(\bar{x}, \bar{y}, \bar{z})$ and (x, y, z) are the co-ordinates of P̄ and P respectively. Eliminating these co-ordinates and substituting in expression (2.4) Hopkins obtained

$$W=dW\sum_{i=1}^{k}=$$

$$\sum_{i=1}^{k}\Delta_i\left[n\,.\,\frac{(L+\bar{L})(X-\bar{X})+(M+\bar{M})(Y-\bar{Y})+(N+\bar{N})(Z-\bar{Z})}{1{\cdot}0+L\bar{L}+M\bar{M}+N\bar{N}}\right]$$

on summing to the kth surface.

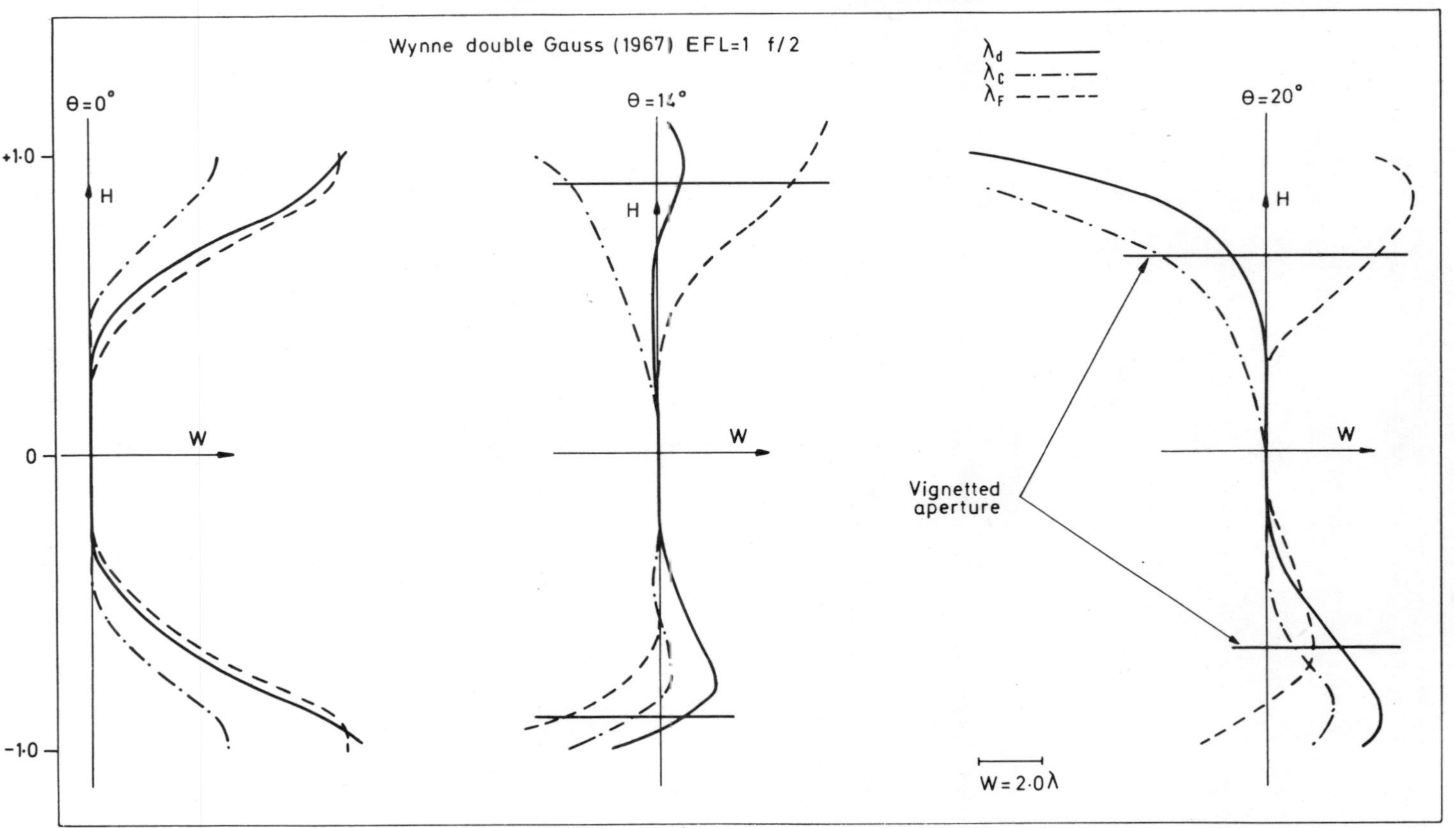

FIG. 21(*a*). Wavefront aberration in the meridian section for the Wynne double Gauss design.

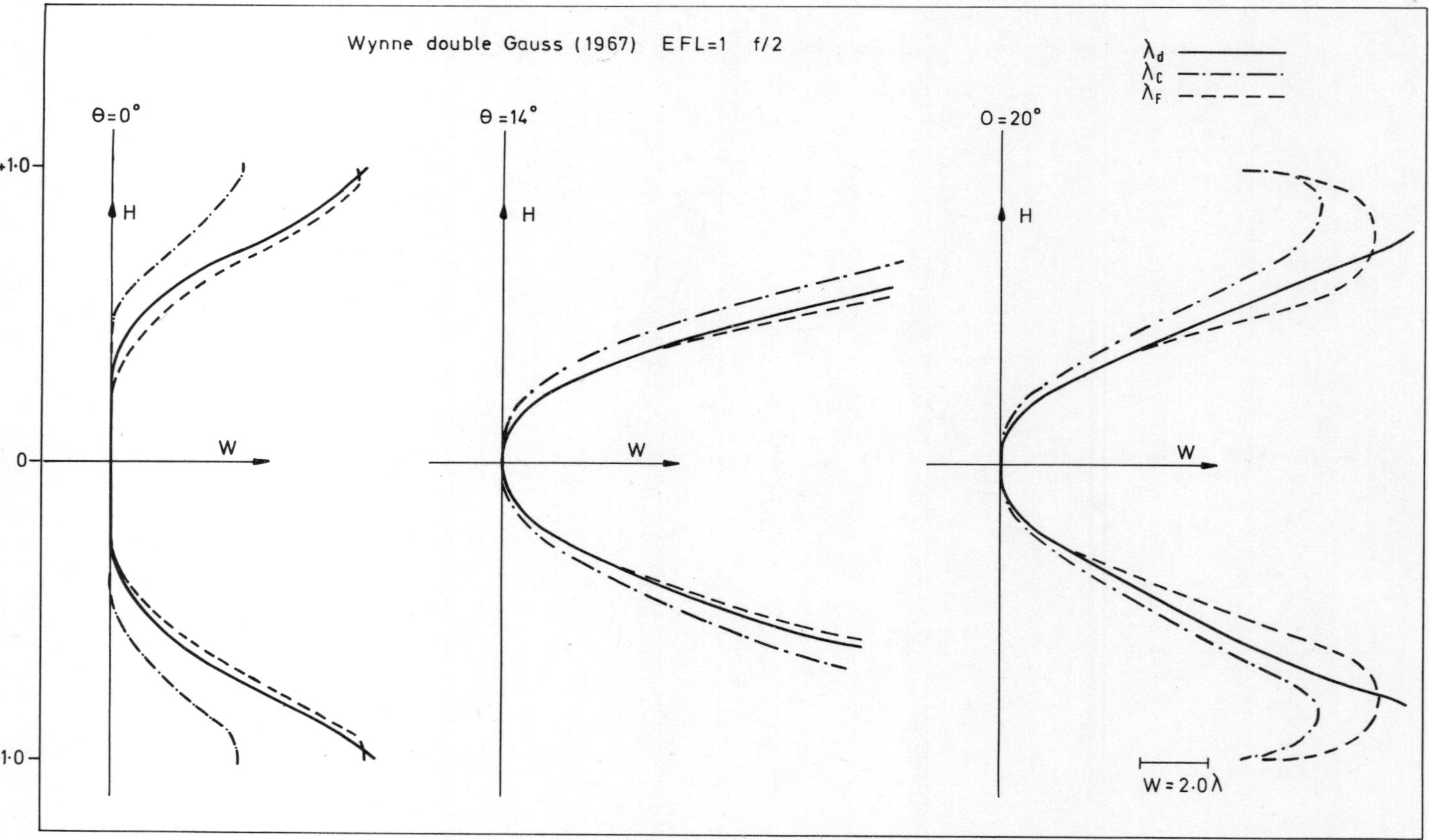

FIG. 21(*b*). Wavefront aberration in the sagittal section for the Wynne double Gauss design.

These equations apply equally to aspheric refracting surfaces, as the expressions depend only on the co-ordinates of the two rays at the current surface. The wavefront aberration of the whole system is obtained by summing the surface contributions Δ_i. The aberration of the finally emergent wave is thereby obtained, referred to the invariant focus.

In the final image space of the optical system, it is desirable to refer the aberration of all points on the wavefront to a common focus, for example, to the point of intersection of the principal ray with the Gaussian image plane. For this purpose, the position of the wavefront must be specified, since the new reference point will not necessarily be an invariant focus. In general, the wavefront will be taken to be located in the exit pupil, and a focal shift term has to be added to the aberration. The exact form of the expression for the focal shift terms depends on the position of the image and exit pupil, special forms being required for infinite object conjugates and telecentric stops. The form of these shift equations are to be found in Hopkins' paper (1952). The wavefront aberration for the double Gauss example (Fig. 21) was calculated by this method. The wavefront aberration in the meridian section is shown plotted as the abscissa and the normalized pupil height as the ordinate.

2.5.2. *Wavefront aberration and ray aberration—a further method of calculation*

Conrady (1960) showed the relationship between wavefront aberration and spherical aberration, and Nijboer (1942) and Hopkins (1950) have treated the more general case of the relationship between wavefront aberration and ray aberration for extra-axial images and all pupil azimuths.

Let $\delta\xi'$ and $\delta\eta'$ be the components of transverse ray aberration measured in the image plane as a displacement from the intersection with the principal ray. Further, let R be the radius of the reference sphere, n' the refractive index of the medium in the image space, and α the semi-aperture angle following the convention of Fig. 12. Then

$$\delta\xi' = -\frac{R}{n'}\frac{\partial W}{\partial x} = \frac{1}{n' \sin \alpha}\frac{\partial W}{\partial x}$$

$$\delta\eta' = -\frac{R}{n'}\frac{\partial W}{\partial y} = \frac{1}{n' \sin \alpha}\frac{\partial W}{\partial y}$$

or in polar co-ordinates

$$\delta\xi' = -\frac{1}{n' \sin \alpha}\left(\sin \phi \frac{\partial W}{\partial \rho} + \frac{\cos \phi}{\rho}\frac{\partial W}{\partial \phi}\right)$$

$$\delta\eta' = -\frac{1}{n' \sin \alpha}\left(\cos \phi \frac{\partial W}{\partial \rho} + \frac{\sin \phi}{\rho}\frac{\partial W}{\partial \phi}\right)$$

where ρ and ϕ are the pupil co-ordinates.

These are approximate relationships assuming that the angular aberration χ is small, and the radius R of the reference sphere is great compared to the wavelength of the light. Exact relationships have been obtained by Rayces (1964).

From the above equations, it follows that if the values of $\delta\xi'$ and $\delta\eta'$ are known for rays in the section of azimuth ϕ, the wave-front aberration at the point (ρ, ϕ) is

$$W = -\frac{n'}{R}\int_0^\rho (\delta\xi' \sin \phi + \delta\eta' \cos \phi)\, d\rho$$

where ϕ is constant during the integration. If the image is at infinity, one uses the angular aberrations χ_x and χ_y, where

$$\chi_x = -\frac{\delta\xi'}{R}$$

and

$$\chi_y = -\frac{\delta\eta'}{R}$$

Then

$$W = n'\int_0^\rho (\chi_x \sin \phi + \chi_y \cos \phi)\, d\rho$$

We may obtain the expressions in terms of the semi-aperture angle a using the substitution $\rho = R \sin a$; since on differentiating

$$d\rho = R \cos \alpha \, d\alpha$$

whence

$$W = -n'\int_0^a (\delta\xi' \sin\phi + \delta\eta' \cos\phi) \cos\alpha \, d\alpha$$

The evaluation of these integrals allows the wavefront aberration to be determined from the geometrical aberrations obtained by ray tracing.

To illustrate the method, we shall consider the case of an axial image. From the symmetry the aberration is independent of ϕ, so we need consider only the rays traced in one plane, say the $Y-Z$ plane taken as the meridian section. Let L' be the longitudinal spherical aberration, and U' be the finite convergence angle of the ray. Then

$$\delta\xi' = 0 \qquad \text{and} \qquad \delta\eta' = -\delta L' \tan U'$$

and so

$$W = n'\int_0^a \delta L' \tan U' \cos\alpha \, d\alpha$$

from which we obtain

$$W = n'\int_0^a \delta L' \sin U' \, dU'$$

by allowing that α is approximately equal to U' for cases of small angular aberration. The integration can be performed graphically by plotting values of $\delta L' \sin U'$ against U', i.e. a transverse aberration plot. The value of W for the aperture α is then the area between the curve, the U'-axis and the ordinate at $U' = \alpha$. This can easily be evaluated by counting squares.

2.5.3. *Conrady $(d-D)$ chromatic aberration*

Another of Conrady's contributions to the art of obtaining aberration data with the greatest economy of calculation, is the $(d-D)$ method of calculating chromatic aberration as an optical path difference.

Conrady (1960) originally formulated this method as a condition for axial achromatism, but also considered the more general case of a finite amount of chromatic aberration. The condition for achromatism between wavelength λ and $\lambda + d\lambda$ is

$$\sum_{i=1}^{k}(d_i - D_i)\delta n = 0$$

where d_i is the axial separation between $(i-1)$th and ith surfaces, D_i the path along the finite ray (usually marginal ray), and δn the dispersion for the wavelength range λ to $\lambda + d\lambda$ in a material of refractive index n at wavelength λ. D_i may be calculated as described in § 2.4.2.

In the case of imperfect achromatism the chromatic aberration is

$$\text{Chr.} = \sum_{i=1}^{k}(d_i - D_i)\delta n;$$

being positive for undercorrection. It is not difficult to obtain a general form for extra-axial images. (See for example Hopkins, 1950.)

Let Σ be a wavefront associated with a disturbance of wavelength λ. It emerges as Σ'_λ after transmission by an optical system of k surfaces. In Fig. 22 let ray a be a principal ray, and ray b be a typical skew ray for the corresponding extra-axial image point. At the wavelength $\lambda + d\lambda$, the wavelength difference $d\lambda$ is sufficiently small for the ray neighbouring b (shown as broken lines) to have optical paths equal to that of a to a first approximation. This follows from Fermat's principle, i.e. that for ray paths close to the physical ray, the optical pathlength has a stationary value. Let the wavefronts Σ'_λ and $\Sigma'_{\lambda+d\lambda}$ coincide at the principal ray. Then the chromatic aberration for the

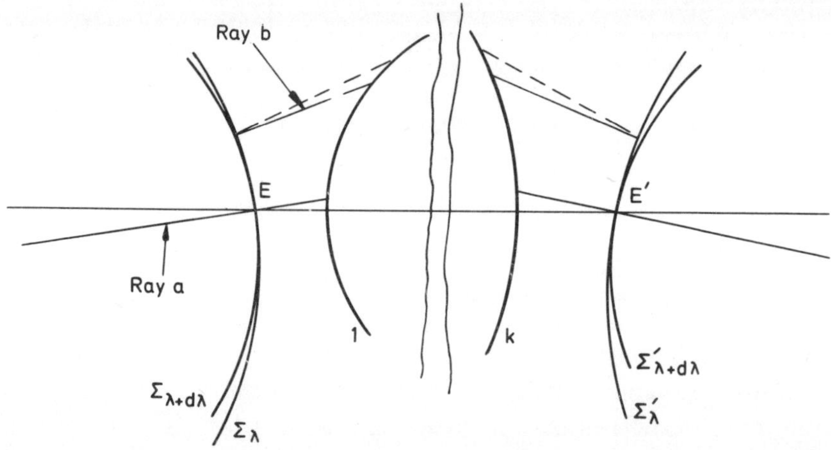

FIG. 22. The Conrady $(d - D)$ chromatic aberration for skew rays.

wavelength $\lambda+d\lambda$ is defined by the deviation of the wavefront $\Sigma'_{\lambda+d\lambda}$ from Σ'_λ. The chromatic aberration

$$\delta W'_{d\lambda}=\delta W'_{\lambda+d\lambda}-\delta W'_\lambda$$

again taken positive if the wavefront $\Sigma_{\lambda+d\lambda}$ is in front of Σ_λ. Now the optical path difference

$$\delta W'_\lambda=(\bar{D}-D)n$$

is the wavefront aberration at wavelength λ, where $\bar{D}$ is the path along the principal ray, and the chromatic aberration is then:

$$W'_{\lambda+d\lambda}-W'_\lambda=(\bar{D}-D)\,\delta n+(\delta\bar{D}-\delta D)n+(\delta\bar{D}+\delta D)\,\delta n$$

But the second term represents the change in optical paths between neighbouring rays which we allow to be zero. The third term is a second order small quantity and can also be neglected. Hence

$$W'_{d\lambda}=\sum_{i=1}^{k}\delta W'_{d\lambda}=\sum_{i=1}^{k}(\bar{D}-D_i)\,\delta n \quad \text{on summing to the } k\text{th surface}$$

Since the dispersion of the air is effectively zero, the elements of this summation have only to be evaluated for the paths through glass. In the axial case, $\bar{D}=d$ and the equation reduces to

$$W'_{d\lambda}=\sum_{i=1}^{k}(d_i-D_i)\,\delta n$$

We can use the quantity $W'_{d\lambda}$ to obtain the approximate wavefront aberration for wavelengths other than the principal wavelength without actually evaluating the equations of §§ 2.5.1 and 2.5.2. Once the wavefront aberration of the principal wavelength has been obtained, we simply take the sum $W'_\lambda+\delta W'_\lambda$ to obtain the required wavefront aberration. This can result in a considerable saving of computation.

If the wavefront aberration has been calculated using the Hopkins' method described in § 2.5.1, then the optical path-length chromatic aberration can be obtained very easily;

$$\delta W'_{d\lambda}=dW'_{\lambda+d\lambda}-dW'_\lambda$$

or $$\delta W'_{d\lambda}=e\,\delta n \quad \text{with the former notation}$$

and hence on substituting and summing to the kth surface

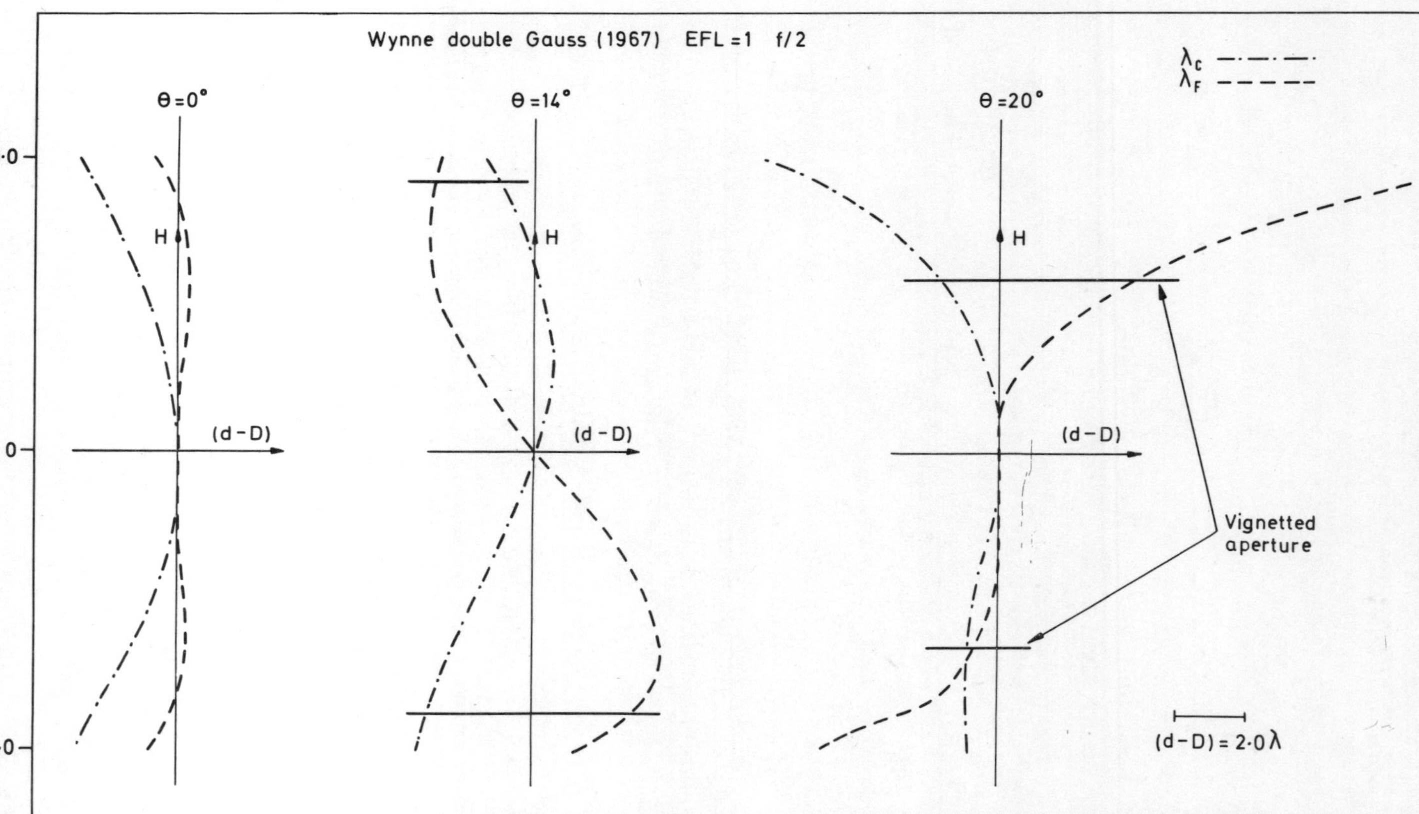

FIG. 23. The Conrady $(d - D)$ chromatic aberration in the meridian section for the Wynne double Gauss design.

$$W_{d\lambda}=\sum_{i=1}^{k}\delta W_{d\lambda}=$$
$$\sum_{i=1}^{k}\Delta_i\left[\delta n\frac{(L+\bar{L})(X-\bar{X})+(M+\bar{M})(Y-\bar{Y})+(N+\bar{N})(Z-\bar{Z})}{1\cdot 0+L\bar{L}+M\bar{M}+N\bar{N}}\right]$$

There is no focal shift term since it corresponds to a second order small quantity.

Graphically, we may either plot the chromatic optical path difference $W'_{d\lambda}$ as abscissa with the normalized pupil height as ordinate, or, alternatively, use $W_{d\lambda}'$ to calculate the total wavefront aberration for each wavelength. The former method more clearly indicates the chromatic variation of primary aberration, and so is shown in Fig. 23 for the double Gauss example. The wavefront aberration in Fig. 22 was obtained from ray tracing at three wavelengths and so provides a comparison with the Conrady approximation.

3

Aberration Tolerances

In this chapter, it is proposed to show how the aberration data obtained by the methods so far described can be used to describe the performance of the objective in a more concise form. There are several reasons for requiring such criteria. It seems to be clearly established by experience that it is impossible to reduce the aberrations precisely to zero in any but trivial optical systems, although this has not yet been rigorously proved. It is, therefore, essential to establish tolerances for aberration, so that the optical designer may decide when a design is good enough for its purpose or which of two systems is the better. Such a tolerance system can also help the designer to put manufacturing tolerances to the optical and mechanical components of a system. The tolerances must also permit the non-specialists in optics, such as the electronic engineer or mechanical designer, to evaluate the influence of the optical system on the instrument or system as a whole, since it is not everyone who shares the optical designer's intuitive feeling for such quantities as OSC′ or for the information contained in an interferogram. Naturally, these requirements are not easily met, and indeed this aspect of optical design is currently the subject of much discussion and controversy.

Most aberration tolerances are derived from scalar diffraction theory, but one important group of tolerances is generally calculated from geometrical optics using ray tracing methods. This group consists of figure of merit functions, which are used mainly in lens optimization programs.

3.1. FIGURE OF MERIT FUNCTIONS

The figure of merit function, a form of tolerance which might less optimistically be called a figure of demerit or defect function, is used in lens optimization programs to indicate the progress of the correction process. These programs work by an iterative process of varying the lens parameters, determining the change in aberration, and calculating a figure of merit, and then, by appropriate choice of parameter increment, the program attempts to minimize the overall aberration. If the correction process has broken down, the figure of merit begins to increase and indicates that some intervention in the process is necessary.

An important feature of these functions is that they should be easily calculated from the minimum of ray tracing. The usual method is to take the mean square of the residual aberration (transverse ray or wavefront) after each cycle of correction. This mean is

$$\bar{A}^2 = \frac{1}{M} \sum_m A_m^2$$

where A_m is the aberration of the rays $m = 1, 2, \ldots, M$. If it is known which ray represents the aberration most detrimental to performance, or if a mixture of aberration types are included, it is possible to take a weighted mean:

$$\bar{A}^2 = \frac{1}{M} \sum_m (wA)_m^2$$

where w represents weighting factors.

As an example we may consider the radius of gyration merit function calculated from transverse ray aberrations:

$$\bar{A}^2 = \frac{1}{M} \sum_{i=1}^{M} (x_i^2 + y_i^2)$$

Unfortunately when the image spot has a dense centre and a broad halo of rays, this tolerance fails, as the rays in the halo contribute little to the image defect at high spatial frequencies, but are a very significant part of the merit function.

In an attempt to overcome this sort of problem, many different functions have been developed (see, for example, *Proceedings of the Conference on Lens Design with Large Computors*, held at the Institute of Optics, Rochester, 1966). It is then a matter of pride to be able to show a graph of merit function versus cycles of integration where the slope is very steep and only tails off at the near-perfect system. Unfortunately, while it is not too difficult to obtain a steep slope, it is very difficult to devise a figure of merit function which represents a realistic measure of lens quality.

3.2. THE RAYLEIGH LIMIT

The Rayleigh limit is a tolerance applicable to systems which have almost perfect correction for certain aberrations; in other words, the image-forming wavefronts closely approximate to the reference sphere, that is, a spherical wavefront.

A tolerance of this kind was first suggested by Lord Rayleigh (1879). His suggestion was based on a study of the special case of defocus as an aberration, primarily for use with astronomical telescopes. The tolerance relies on the fact that provided the wavefront at the exit pupil deviates from a reference sphere centred on the ideal image point by less than a quarter of a wavelength, then the effect upon the Airy pattern is minimal. The Airy pattern is the light distribution formed on the image plane by an aberration-free system having a circular pupil, from a point object. [Born and Wolf (1964) give an excellent treatment of the theory of diffraction formation of images.] Consequently, a system is said to be 'Rayleigh limit corrected' if the wavefront deformation is less than a quarter of a wavelength of light.

It can be shown that the geometrical wavefront calculated by the methods described in § 2.5.1 is an accurate representation of the true wavefront found by application of scalar diffraction theory, subject to some restrictions (Born and Wolf, 1964). These restrictions need not concern us provided the wavefront is many wavelengths away from the focus and the aperture is large compared with a wavelength of light. The tolerance can therefore be established by inspecting a display of the form described in § 2.5.1 or § 4.2, thus providing a simple rule-of-thumb interpretation of performance.

3.3. STREHL INTENSITY RATIO (*Definitionshelligheit*)

It is found that the kind of aberration present in a system does have some influence on the magnitude of a diffraction-based criterion such as the Rayleigh limit, though, for a high-quality system, the general effect is similar, namely, a drop in the intensity of the central maximum of the Airy pattern together with the spreading of the light into other regions of the pattern.

The lowered central intensity, expressed as a fraction of the original, is known in English-speaking countries as the 'Strehl intensity ratio', named after K. Strehl who suggested it in 1894. A Strehl intensity of 0·8 is generally accepted as the tolerance limit for a change of focus and is equivalent to the Rayleigh quarter-wavelength criterion. Maréchal (1947) has shown that the decrease in the Strehl ratio resulting from the presence of aberration is determined by the total variance E of the wavefront aberration W.

Let us define

$$E = \overline{W^2} - (\overline{W})^2 \tag{3.1}$$

where the bar denotes a mean value over the pupil domain and therefore

$$\overline{W^2} = \frac{I_0}{S}\iint_{\text{pupil}} W^2 \, dS \tag{3.2}$$

and

$$\overline{W} = \frac{I_0}{S}\iint_{\text{pupil}} W \, dS \tag{3.3}$$

where I_0 is the intensity at the origin, S the pupil area, and dS a pupil element. Maréchal actually used a slightly different definition for the total variance, but, in principle, he showed that, for small aberration, the Strehl intensity I is given by

$$I = 1{\cdot}0 - \frac{4\pi^2}{\lambda^2} E \tag{3.4}$$

In effect, the Maréchal technique amounts to evaluating the diffraction integral for the intensity distribution of a point object subject to certain approximations. These approximations can be justified for small amounts of aberration.

Before considering the problem of evaluating the Strehl intensity for particular designs, it is useful to consider how the intensity ratio may be developed into a tolerance for the maximum possible aberration permissible in an optical system.

Let us suppose we may tolerate a 20 per cent drop in the central intensity. Putting $I \geqq 0{\cdot}8$, we obtain

$$\frac{4\pi^2}{\lambda^2}E \leqq 0{\cdot}2$$

$$E \leqq \frac{1}{197\lambda^2}$$

This implies the root-mean-square σ of the wavefront aberration is given by

$$\sigma \leqq \frac{1}{14\lambda}$$

which is equivalent to requiring that the root-mean-square departure of the wavefront from a reference sphere centred on the diffraction focus shall not exceed $\lambda/14$.

We can now use this result to obtain tolerances for various terms of wavefront aberration function. Taking the simple case of defocus for our example, we have the aberration function W for the defect of focus ζ given by

$$W = \frac{(x^2 + y^2)\zeta}{2R^2}$$

measured from the reference sphere of radius R, and taken at a point having co-ordinates (x, y).

Now, if we transpose to normalized polar co-ordinates for a circular pupil, we have

$$W = \frac{\rho^2}{2}\left(\frac{a^2}{R^2}\right)\zeta$$

where ρ is the pupil parameter having a maximum value of ± 1 for a pupil diameter a. Also, we put $W = A_{20}\rho^2$ in order to simplify the substitution and then using equations (3.2) and (3.3) in equation (3.4) we obtain

$$E = \frac{A_{20}^2}{\pi}\int_0^{2\pi}\int_0^1 \rho^4 \,.\, \rho d\rho d\phi - \left(\frac{A_{20}}{\pi}\int_0^{2\pi}\int_0^1 \rho^2 \,.\, \rho d\rho d\phi\right)^2$$

$$E = 2A_{20}^2\int_0^1 \rho^5 d\rho - \left(2A_{20}\int_0^1 \rho^3 d\rho\right)^2$$

$$E = \frac{A_{20}^2}{3} - \frac{A_{20}^2}{4} = \frac{A_{20}^2}{12}$$

Now applying the tolerance $\sigma \leqq 1/14\lambda$, we have

$$\frac{A_{20}}{\sqrt{12}} \leqq \frac{1}{14\lambda}$$

Hence

$$W \leqq 0{\cdot}25\lambda\rho^2$$

and

$$\zeta \leqq \pm \frac{0{\cdot}5\lambda}{(a/R)^2}$$

which is the well-known result for focal defect as originally obtained by Lord Rayleigh.

The general expression for the aberration function in terms of polar co-ordinates can be written as the polynominal expression

$$\begin{aligned} W(\rho\phi\eta) = {} & A_{020}\rho^2 + A_{111}\eta\rho\cos\phi \\ & + A_{040}\rho^4 + A_{130}\eta\rho^3\cos\phi + A_{222}\eta^2\rho^2\cos^2\phi \\ & + A_{220}\eta^2\rho^2 + A_{311}\eta^3\rho\cos\phi \\ & + A_{060}\rho^6 + \ldots \end{aligned}$$

allowing for terms in ξ to be zero by considering only object points with finite values of the co-ordinate η; all other object points may be obtained by rotating a centred optical system.

The terms of the series represent the various orders of aberration. Terms A_{020} and A_{111} are longitudinal and transverse defocus respectively; terms A_{040}, A_{130}, A_{222}, A_{220} and A_{311} are the five primary aberrations; and A_{060} and higher terms represent the high order aberration of the pupil and field. We may draw up a set of tolerances for the terms influencing point imagery by calculations of the type illustrated above for the case of longitudinal defocus. It is necessary in each calculation to include sufficient terms to allow the condition of best focus to be obtained. For example, in a lens with significant amounts of secondary spherical aberration, the best focus is obtained only by the correct choice of the coefficients A_{020}, A_{040} and A_{060}.

The results are tabulated in Table 3.1.

TABLE 3.1

Aberration	Aberration coefficient	Coefficient tolerance
Longitudinal defocus	A_{020}	$A_{020} = \pm 0{\cdot}25\lambda$
Primary spherical	A_{040}	$A_{020} = \pm 0{\cdot}95\lambda$ $A_{040} = \mp 0{\cdot}95\lambda$
Secondary spherical	A_{060}	$A_{020} = \pm 2{\cdot}24\lambda$ $A_{040} = \mp 5{\cdot}62\lambda$ $A_{060} = \pm 3{\cdot}74\lambda$
Primary coma	A_{130}	$A_{111} = \pm 0{\cdot}4\lambda$ $A_{130} = \mp 0{\cdot}6\lambda$
Primary astigmatism	A_{222}	$A_{020} = \pm 0{\cdot}175\lambda$ $A_{222} = \mp 0{\cdot}35\lambda$

A Strehl intensity type of tolerance has been obtained by Welford (1963a) for imagery of a line by a rectangular pupil, as in a spectroscopic instrument. Hopkins (1966) has shown how the Strehl intensity can be calculated from ray tracing data. Essentially, the wavefront polynomial is determined by tracing the same number of rays as the terms in the polynomial. This polynomial will have the form

$$W = \sum_n \sum_q W(nq)(x^2 + y^2)(n - q/2)y^q$$

for an extra-axial image with the co-ordinates $(x_j\, y_j)$. This equation may be inverted to give

$$W(n, q) = \sum_j k(n, q, j) W_j$$

Then, for a particular wavefront polynomial $W(x, y)$, the total variance E will have the form

$$E = \sum_i \sum_j p(i, j) W_i W_j$$

where $p(i, j)$ are the coefficients of a new polynomial.

As Hopkins' method is intended to be used with a small number of rays, the variance has an approximate value only. In principle, the method could be extended to cover the case where the wavefront polynomial has been determined with much higher precision. Hopkins shows that it is possible to use this criterion in automatic optimization procedure for designing systems such as microscope objectives and telescope objectives. It is customary to say that these systems are capable of being 'diffraction limited', that is, having a Strehl intensity exceeding 0·8 or thereabouts. The double Gauss example does not fall into this category, so the Strehl intensity ratio has no significance and was therefore not evaluated.

3.4. OPTICAL TRANSFER FUNCTION (OTF)

There is an important range of optical systems for which the residual aberrations are sufficiently large that the Strehl intensity ratio becomes an invalid criterion of image quality. This range includes photographic, television and similar optics where the light gathering capability is a more significant quality than the ultimate in resolution. H. H. Hopkins (1966) has suggested the equation given by Maréchal

$$I = 1{\cdot}0 - \frac{4\pi^2}{\lambda^2} E$$

is valid for values of the variance E that lead to values of I greater than 0·7. We have seen that, for simple cases of wavefront aberration, the Strehl ratio of 0·8 roughly corresponds to the Rayleigh limit of a quarter-wavelength departure from a spherical wavefront.

Clearly, we need a new tolerance to include systems which are far from this class of 'diffraction-limited' systems. The optical transfer function has been developed to deal with just such optical systems, though, fortunately, it is by no means restricted to these cases alone. Another way to characterize these optical systems is to say that the aperture × field product is large compared with other optical systems. By this we imply that the high aperture or the large field coverage may contribute to the departure from diffraction-limited performance. The optical transfer approach to imagery has the further virtue of complete applicability to extended image fields found in this class of systems. The Strehl intensity ratio, in marked contrast, is limited to point imagery or, at most, to image fields sufficiently small that the aberration is substantially constant over the area surrounding the point image.

3.4.1 *Theoretical considerations*

The theoretical foundations that underline the calculation of the optical transfer function, originated by P. M. Duffienx and

O. Shade and further developed by Hopkins (1953, 1955a, etc.) and others, is beyond the scope of this monograph but is given by several authors, for example, Miyamoto (1961), Born and Wolf (1964), and Bray (1965).

The more important facts becoming evident from a study of the basic theory, seen from the point of view of the application of this form of tolerance, concern the general conditions of validity. These relate principally to the nature of the illuminations and the condition of isoplanatism.

Physicists find that, as a consequence of extending the theory of image formation to include objects of finite extent, it becomes necessary to consider the coherence properties of adjacent object points. If light from any two such points is brought into superposition and no interference fringes result, then the illumination is said to be incoherent. Under these conditions the optical transfer function takes the form which has found the most general application and so will be discussed most fully in the following sections.

The concept of a transfer function is not limited to incoherent illumination. At the other extreme, coherent illumination, when the interference fringes have unit contract according to the Michelson definition, it can be shown that the frequency response for the system is best treated as the transfer of amplitude from the object to the image plane. In this case, the transfer function can be identified with the pupil function of the objective (see Born and Wolf, 1964, p. 481). A valuable characteristic of coherently illuminated optical systems now becomes apparent because it follows that manipulation of the pupil function by, say, masking allows very considerable control of the image plane amplitude distribution in a direct way. We may anticipate, therefore, that as coherent optical systems become more widely used the specification of these systems by their transfer function will assume importance.

The general case of partially coherent illumination is more difficult to treat analytically because it is necessary to know the form of the object intensity distribution and the nature of the source geometry in order to calculate the degree of partial coherence. Hopkins (1953) has treated the image formation of periodic line structures in partially coherent light, and shown that for the specification of this case a set of cross transfer factors are required for each pair of spatial frequencies arising from the pupil and the source. For two-dimensional objects there are factors involving each frequency pair in the two directions. This is significantly different to the incoherent case where there is a single response factor for each spatial frequency.

Steel (1957) has analyzed the effect of small aberrations on the images of a partially coherent object having the form of periodic lines and single lines. By restricting the source intensity distribution to centred circular symmetry, he was successful in developing a set of tolerances based on the optical transfer function. His tolerances represent an extension to partially coherent illumination of the Maréchal tolerances described in §3.3. In this case the tolerance condition is determined by the reduction of the response function to a value 0·8 of the original ideal magnitude. As before, certain approximations are necessary in the calculations and they limit the application of the tolerances. In practice these limitations are found to be rather severe. This is some indication of the intractability of calculations in partially coherent illumination, since, even with the restricted circumstances treated, the tolerances prove to bc accurate only for low contrast objects and small reductions in image contrast.

In developing the theory of imagery by optical transfer it is necessary to assume that the condition of isoplanatism holds true. Essentially, this implies the aberration function of the pupil should be a slowly varying function in the spatial frequency space over the range covered by a few periods of the spatial frequency of interest. Dumontet (1955) discusses the implications of isoplanatism in detail. We may take it that for a well-corrected optical system this condition is substantially satisfied. For condenser systems, however, this may not be true.

3.4.2. *The optical transfer function of the incoherently illuminated lens*

The concept of the limit of resolution as expounded by Lord Rayleigh leads us to realize that the wave nature of light imposes a fundamental limitation on the image forming capability of a lens of any given aperture. We discover that this same limitation has an important meaning in the optical transfer treatment of image formation. Broadly, the lens is not capable of transferring an intensity distribution from one plane (the object) to another (the image) with complete fidelity of contrast or position (apart from geometrical distortion). In fact we find the frequency response progressively decays to a zero value at a finite frequency, generally called 'the cut-off frequency'. Stated mathematically, the response of the optical system to line frequencies in incoherent illumination is the Fourier transform of the point spread function, or, in an alternative formulation, the frequency response is the autocorrelation of the pupil function. Let us represent the frequency response by the function $\mathscr{L}(u_0, v_0)$ at the orthogonal frequencies u_0 and v_0; then by the second definition

$$\mathscr{L}(u_0, v_0) = \iint_{\text{pupil}} k\left(u + \frac{u_0}{2}, v + \frac{v_0}{2}\right) k^*\left(u - \frac{u_0}{2}, v - \frac{v_0}{2}\right) du\, dv$$

where u and v are the normalized pupil co-ordinates, and the pupil function is given by

$$k(u, v) = \exp\left[-\frac{2\pi i}{\lambda} W(u, v)\right]$$

Hence, on substituting for $k(u, v)$ we obtain

$$\mathscr{L}(u_0, v_0) = \iint_{\text{pupil}} \exp\left\{-\frac{2\pi i}{\lambda}\left[W\left(u + \frac{u_0}{2}, v + \frac{v_0}{2}\right) - W\left(u - \frac{u_0}{2}, v - \frac{v_0}{2}\right)\right]\right\} du\, dv$$

From this equation it follows that we need only discover the form of the wave aberration function $W(u, v)$ in order to be in the position to evaluate the response function of any optical system. It is for this reason that the application of Fourier analysis techniques to optics has proved to be such a powerful tool. The actual evaluation of image intensity distributions in the presence of particular amounts of aberration is, in principle at least, easily accomplished using numerical integration with fast electronic computers.

The physical interpretation of the operation of the autocorrelation function was shown by Hopkins (1953) to be equivalent to shearing the pupil with itself by an amount corresponding to the spatial frequency of interest and then determining the integrated intensity transmitted by the pupil. The resulting sheared pupils will only transmit over the region common to both pupils and this determines the domain of integration of the above equation. Usually the imaging of off-axis points is accompanied by vignetting of the pupil and so the shape of the vignetted pupil must be determined before the numerical integration can be completed. (One of the difficulties alluded to in § 3.4.1 concerning the transfer function in partially coherent illumination arises because, in addition to shearing the pupils, it is necessary to superimpose the region of the equivalent source in the domain of integration and allow for the resulting cross-coefficients.)

3.4.3. *The optical transfer function of the perfect lens*

The importance of knowing the magnitude of the diffraction-limited response becomes evident when specifying the performance requirement of an optical system. Clearly, it is absurd to demand a response in excess of the diffraction limit, though the temptation to do so may be strong in, say, an electro-optic system where even the response of the perfect optical system may have a considerable degradation at comparatively low frequencies as a consequence of low aperture or long wavelength. Further, we may wish to set a tolerance on the departure from the perfect response function. The response of the aberration-free lens is dependent on the common area of the lens pupils after the appropriate shearing operation. Hopkins (1953) has given a general expression for the response of an aberration-free lens to line frequencies. When such a system is defocused, the pupil function takes a simple form depending only on the coefficient of the defect of focus. Hopkins (1955a) examined the effect of defect of focus on the frequency response of the aberration-free lens. He showed the response function can be transformed to an integral in transcendental functions, which can be evaluated numerically by expanding in terms of Bessel functions J_1, J_2, etc. The pupil function has the form

$$k(u, v) = \exp\left[\frac{2\pi i A_{20}}{\lambda}(x^2 + y^2)\right] \quad \text{within the pupil}$$

$$k(u, v) = 0 \quad \text{outside the pupil}$$

for a defect of focus giving a coefficient of wave aberration A_{20} at a wavelength λ. Then, if the substitution $y = \sin\theta$ is made, the one-dimensional response function becomes

$$\mathscr{L}(u, 0) = \frac{4}{\pi a} \cos \tfrac{1}{2}a\,|u| \int_0^\beta \sin(a \cos\theta) \cos\theta \, d\theta - \frac{4}{\pi a} \sin \tfrac{1}{2}a\,|u| \int_0^\beta \cos(a\cos\theta)\cos\theta\, d\theta$$

where $a = (4\pi/\lambda)A_{20}\,|u|$ with the limit of integration $\beta = \arccos \frac{1}{2}|u|$. For numerical evaluation we have the convergent series form

$$\begin{aligned}\mathscr{L}(u, 0) = \frac{4}{\pi a} \cos \tfrac{1}{2}a\,|u| \,\{&\beta J_1(a) + \tfrac{1}{2} \sin 2\beta\,[J_1(a) - J_3(a)] \\ &- \tfrac{1}{4}\sin 4\beta\,[J_3(a) - J_5(a)] + \ldots\} \\ - \frac{4}{\pi a} \sin \tfrac{1}{2}a\,|u|\,\{&\sin\beta\,[J_0(a) - J_2(a)] \\ &- \tfrac{1}{3}\sin 3\beta\,[J_2(a) - J_4(a)] \\ &+ \tfrac{1}{5}\sin 5\beta\,[J_4(a) - J_6(a)] - \ldots\}\end{aligned}$$

This expression reduces to

$$\mathscr{L}(u, 0) = \frac{2}{\pi} \arccos \tfrac{1}{2}\,|u| - \frac{u}{\pi}\sqrt{\left(1 - \frac{u^2}{4}\right)}$$

at perfect focus for a circular pupil. The restriction to line frequencies implies no loss of generality, because, as Hopkins shows, by a suitable rotation ψ of the pupil axes and corresponding aberration function axes any pair of frequencies (u_0, v_0) can be reduced to a single frequency $(u, 0)$, the equation of rotation being $\tan\psi = v_0/u_0$. The response function for the double Gauss lens compared with the response of the corresponding perfect lens is shown in Fig. 24(*a*). The normalized spatial frequencies u_0 and v_0 can be derived from the geometrical quantities of the optical system by the equations

$$u_0 = \frac{\lambda}{n' \sin a'} \cdot f$$

$$v_0 = \frac{\lambda}{n' \sin a'} \cdot g$$

noting that spatial frequencies are obtained from co-ordinate distances by the reciprocal relationship

$$f=\frac{1}{\xi'} \quad \text{and} \quad g=\frac{1}{\eta'}$$

where n' sin α' is the numerical aperture and λ the wavelength of the illumination.

The transfer function for a rectangular aperture is of interest in the design of spectrographs. Let the angular width of the side of the aperture perpendicular to the direction of the line structure be α' as seen from the axial image point, then the response function for perfect focus is

$$\begin{aligned} \mathscr{L}(u) &= 1-\tfrac{1}{2}\,|u| \qquad & |u| \leqq 2 \\ &= 0 & |u| > 2 \end{aligned}$$

3.4.4. *The evaluation of optical transfer function*

We saw in § 3.4.2 that from a knowledge of the wave aberration function it is possible to represent the OTF by a double integral taken over the pupil domain. We may determine the aberration function by first ray tracing to obtain the wave aberration at a number of points in the exit pupil, as described in § 2.5, and then fitting a polynomial by the method of least squares. The integration may be performed using a method due to Hopkins (1957), thereby evaluating the response function in a precise way. This method was employed by Goodbody (1958, 1960) in his studies of the influence of aberrations on the response function. Reference to these studies provides valuable numerical data in graphical form. Other methods have been suggested using, for example, Gauss quadrature in conjunction with Legendre polynomials (Barakat, 1962). Unfortunately, while diffraction-based integral procedures are feasible using such numerical integration techniques with fast digital computers, it turns out to be a costly process because of the amount of computer time required to complete the calculation.

Miyamoto (1957a, 1961) has shown that a geometrical approximation to OTF is possible for systems with relatively large amounts of aberration. Miyamoto has treated the monochromatic illumination case while more recently Bray (1965) has extended the theory to include polychromatic illumination. Bray has shown that the computation time for the geometrical approximation is substantially shorter than the rigorous diffraction calculation.

H. H. Hopkins (1955) has discussed the comparative results obtained with the diffraction-based OTF and the geometrical approximation for the special case of defocus in an aberration-free lens. He deduced that, if the magnitude of the wave aberration function is larger than two wavelengths, then both methods of calculation produced results which coincide with each other to an acceptable degree, particularly when taken for the low frequency domain. Similar results have been obtained for astigmatism, spherical aberration and the general case by De (1955), Bromilow (1958) and Miyamoto (1957a, 1961) respectively. The geometrical approximation to OTF now seems to be well established for the lower tolerance systems mentioned previously. The use of geometrical OTF is seemingly justified, particularly bearing in mind the great saving of computer time.

An interesting empirical method has come into use recently for the range of systems for which the wavefront aberration function is of the order of one or two wavelengths, that is, those which fall between the range of validity of the geometrical OTF and the full diffraction-based calculation. It appears that rather better agreement between the two methods of calculation is obtained by taking the product of the geometrical OTF and of the theoretical OTF of a perfect system of the same focal ratio.

Before examining the geometrical OTF let us first consider the nature of the geometrical approximation in optics. Starting from Maxwell's theory of electromagnetic radiation the development of the principles of optics can be seen as a series of simplifying approximations. Maxwell's equations represent the very generalized approach of electromagnetic field theory, so, when we come to apply the boundary conditions of the physical situation, we are forced to place limits on this generality. The first step leads to the formulation of scalar diffraction theory, which is concerned with the propagation of light through scalar waves. The application of the Kirchhoff integral predicts the intensity distributions produced by scalar waves modified by the presence of opaque screens and 'diffracting apertures'. In recent years a major advance in the usefulness of diffraction theory has resulted from the introduction of Fourier techniques and information theory. The transfer function has become a familiar concept in the formation of optical images, in particular the OTF for the response of an incoherently illuminated optical system has been adopted widely as a criterion of lens quality.

The definition of OTF given in the first equation of § 3.4.2 is therefore expressed in terms of physical optics whereas traditionally optics has been concerned with the next stage in the train of the simplifications, namely, the geometrical approximation to imagery. The development of the equations of geometrical optics from the diffraction formulae can be achieved in a logical fashion by applying the mathematical limiting process of allowing the wavelength to tend to zero. Standard texts, for example Born and Wolf (1964), give extensive treatments of the application of this technique but we shall consider the special case of the OTF given by Miyamoto (1957a).

3.4.5. *The calculation of the geometrical optical transfer function*

We start from the definition of the OTF given in § 3.4.2.

$$\mathscr{L}(u_0, v_0) = \iint_{\text{pupil}} \exp\left\{\frac{-2\pi i}{\lambda}\left[W\left(u+\frac{u_0}{2}, v+\frac{v_0}{2}\right) - W\left(u-\frac{u_0}{2}, v-\frac{v_0}{2}\right)\right]\right\} du\, dv$$

and expand the exponential term in the Taylor series

$$-\frac{2\pi i}{\lambda}\left[2\left(\lambda\frac{u_0}{2}\frac{\partial}{\partial u}+\lambda\frac{v_0}{2}\frac{\partial}{\partial v}\right)W+\frac{2}{6}\left(\lambda\frac{u_0}{2}\frac{\partial}{\partial u}+\lambda\frac{v_0}{2}\frac{\partial}{\partial v}\right)^3 W+\ldots + \frac{2}{(2m-1)!}\left(\lambda\frac{u_0}{2}\frac{\partial}{\partial u}+\lambda\frac{v_0}{2}\frac{\partial}{\partial v}\right)^{2m-1} W+\ldots\right]$$

$$= -2\pi i\left(u_0\frac{\partial}{\partial u}+v_0\frac{\partial}{\partial v}\right)W - 2\pi i\frac{\lambda^2}{24}\left(u_0\frac{\partial}{\partial u}+v\frac{\partial}{\partial v}\right)^3 W+\ldots$$

The first term is the only one independent of the wavelength and so all other terms disappear as λ tends to zero. The geometrical approximation can thus be written

$$\mathscr{L}_g(u_0, v_0) = \text{constant}\iint_{\text{pupil}} \exp\left[-\frac{2\pi i}{\lambda}\left(u_0\frac{\partial W}{\partial u}+v_0\frac{\partial W}{\partial v}\right)\right] du\, dv$$

where the value of the constant is chosen to normalize the function for zero spatial frequency to give

$$\mathscr{L}_g(0, 0) = 1{\cdot}0$$

and the quantities $\partial W/\partial u$ and $\partial W/\partial v$ are proportional to the transverse aberration in the chosen image plane.

The same result is obtained by taking the Fourier transform of the geometrical point spread function generated by the spot diagram technique (§ 2.3.5). Let the normalized co-ordinates for the object and image plane be given by

$$X = \frac{a}{\lambda R}\xi'$$

$$Y = \frac{a}{\lambda R}\eta'$$

Then the Fourier transform of the point intensity distribution is

$$\mathscr{L}_g(u_0\ v_0) = \iint_{-\infty}^{\infty} I_g(X,\ Y) \exp\,[2\pi i(u_0 X + v_0 Y)]\ dX\, dY$$

From the theory of spot diagrams we know the geometrical point intensity distribution is given by the Jacobian

$$I_g(\xi', \eta') = n^{-1}\left|\frac{\partial(\xi', \eta')}{\partial(u, v)}\right|^{-1}$$

and hence

$$I_g(X,\ Y) = \frac{a^2}{\lambda^2 R^2} S^{-1}\left|\frac{\partial(u, v)}{\partial(X,\ Y)}\right|$$

where S is the area of the pupil.

Substituting for $I_g(X, Y)$ and noting that the integrand is finite only for the limits defining the lens pupil, we obtain

$$\mathscr{L}_g(u_0, v_0) = \text{constant} \iint_{\text{pupil}} \exp\,[2\pi i(u_0 X + v_0 Y)]\frac{\partial(u, v)}{\partial(X,\ Y)}\, dX\, dY$$

or

$$\mathscr{L}_g(u_0, v_0) = \text{constant} \iint_{\text{pupil}} \exp\,[2\pi i(u_0 X + v_0 Y)]\ du\, dv$$

on employing the properties of the Jacobian operator. Now remembering the normalizing expressions

$$X = \frac{a}{\lambda R}\xi'$$

and

$$u=\frac{x}{a}$$

we obtain from § 2.5.2

$$\delta'\xi=-\frac{R}{n'}\frac{\partial W}{\partial x} \quad \text{and} \quad \delta X=-\frac{1}{\lambda n'}\frac{\partial W}{\partial u}$$

and so on substituting back in the integral, we obtain, as before,

$$\mathscr{L}_g(u_0, v_0)=\text{constant}\iint_{\text{pupil}}\exp\left[-\frac{2\pi i}{n'\lambda}\left(u_0\frac{\partial W}{\partial u}+v_0\frac{\partial W}{\partial v}\right)\right]du\,dv$$

In practice this equation is used to evaluate the one dimensional transfer function employing a numerical summation in place of the integral. The exponential is expanded by Demoivre's theorem giving

$$\mathscr{L}_g(v)=\sum_j\cos\left(-\frac{2\pi}{n'\lambda}\frac{v_0}{}\frac{\partial W_j}{\partial v}\right)+i\sum_j\sin\left(-\frac{2\pi}{n'\lambda}\frac{v_0}{}\frac{\partial W_j}{\partial v}\right)$$

The spatial frequency g_0 in the units of the lens specification is given by

$$g_0=\frac{n'\sin\alpha}{\lambda}v_0$$

and the transverse aberration of the *j*th ray by

$$\partial\eta_j=-\frac{R}{n'}\frac{\partial W_j}{\partial y} \quad \text{or} \quad \delta\eta_j=-\frac{R}{n'a}\frac{\partial W_j}{\partial v_j}$$

The geometrical transfer function in the tangential section is given by

$$\mathscr{L}_g(g_0)=\sum_j\cos\,(2\pi g_0\,\delta\eta_j')+i\sum_j\sin\,(2\pi g_0\,\delta\eta'_j)$$

Suppose we put $\mathscr{L}_g(g_0)=\mathscr{L}^r(g_0)+i\mathscr{L}^i(g_0)$, then we can write

$$M(g_0)=\{[\mathscr{L}^r(g_0)]^2+[\mathscr{L}^i(g_0)]^2\}^{\frac{1}{2}}$$

$$\phi(g_0)=\text{arc tan}\left[\frac{\mathscr{L}^i(g_0)}{\mathscr{L}^r(g_0)}\right]$$

It is found that, when a sinusodial grating of unit contrast is imaged by the optical system, the image contrast reduction is represented by the modulation transfer function $M(g_0)$, and the phase shift $\phi(g_0)$ corresponds to the displacement of the component of spatial frequency g_0. The geometrical transfer function of the sagittal section can be found by scanning the geometrical point intensity distribution in a direction parallel to the ξ axis:

$$\mathscr{L}_g(f_0) = \sum_j \cos(2\pi f_0 \, \delta\xi_j)$$

noting that for a centred system this is a symmetrical function. The majority of lenses are used with sources covering an extended wavelength range and so it is necessary to take account of the variation of the imaging properties with wavelength. No contradiction exists here, for, whilst the geometrical approximation neglects the very rapid oscillations of the electric vector and considers the light energy to be transported along curved paths called rays, it allows for the variation of the paths with wavelength. Consequently the spot diagram for one wavelength is slightly different in magnitude and position to those of all other wavelengths assuming a well-corrected system. To a first approximation we can allow for this effect geometrically by determining the spot diagram for several discrete wavelengths and taking the weighted mean point intensity distribution. The weighting factors $w_j(\lambda)$ are determined according to the relative significance of each wavelength in the spectral energy distribution of the source and the detector. In this way we obtain the geometric approximation to the polychromatic OTF

$$\mathscr{L}_g(g_0) = \sum_j w_j(\lambda) \cos(2\pi g_0 \, \delta\eta_j) + i \sum_j w_j(\lambda) \sin(2\pi g_0 \, \delta\eta')$$

3.4.6 *Graphical presentation of the optical transfer function*

It is clear that a comprehensive description of the imaging capabilities of a lens system can be formed by the use of the OTF. It takes account of the wave-like nature of light, allows for the effect of aberration in point imagery, and in the case of incoherent illumination, at least, can be seen to bare some resemblance to the traditional methods of assessing lens performance, namely, the limiting resolution of bar line charts and similar criteria. A further consideration of some importance when assessing the performance of electro-optical systems stems from the fact that frequency response and frequency filtering are familiar concepts to the users of electronic systems. The frequency response of the overall system can be obtained by simply cascading the individual response functions, provided incoherent coupling is maintained throughout.

Unfortunately, from the viewpoint of presentation, this quality of completeness also implies a substantial amount of information contained within the expression of the function. Therefore, we are faced with the task of reducing the data to a readily understood and easily assimilated form. Ideally, we should prefer a presentation that constitutes an unequivocal measure of the quality of the optical system. In practice, the last requirement is unlikely to be satisfied because the fine balance of the residual aberrations necessary in order to achieve an optimum transfer function is only possible with some foreknowledge of the lens application, such as the most significant spatial frequency. Miyamoto (1957b) has examined this aspect of the transfer function in a study of the aberration balancing of optical systems in combination with their receivers.

The first step we may take towards simplification has been discussed previously in § 3.4.3. This consists of choosing a rotation of the system axes to reduce the two dimensional function $\mathscr{L}(u, v)$ to the sinusoidal line frequency form $\mathscr{L}(u, 0)$. The relationship between the two spatial frequencies u and v and the angle of rotation ψ is given by

$$\psi = \arctan\left(\frac{v}{u}\right)$$

The OTF is a complex and so can be expressed as having a real and an imaginary part:

$$(u) = \mathscr{L}^r(u) + i\mathscr{L}^i(u)$$

The physical quantities concerned are found to be the modulation transfer function $M(u)$ representing the reduction in image contrast and the phase shift $\phi(u)$ representing the displacement of the component of frequency u. From the theory of complex numbers we may write

$$M(u) = \{[\mathscr{L}^r(u)]^2 + [\mathscr{L}^i(u)]^2\}^{\frac{1}{2}}$$

and

$$\phi(u) = \arctan\left[\frac{\mathscr{L}^i(u)}{\mathscr{L}^r(u)}\right]$$

Now, the next step is to display the transfer function graphically. The normal procedure is to combine the two curves of modulation and phase in one plot by using a common abscissa for spatial frequency. The modulation term is conventionally normalized to unity at zero spatial frequency, and the phase takes values from plus to minus π radians. The left-hand ordinate is used for the contrast and the right-hand ordinate for the phase. A plot of this form was adopted to the display of optical transfer calculated by the geometrical approximation for the double Gauss example, and can be seen in Fig. 24. For purposes of comparison, the diffraction-limited performance curve of the equivalent numerical aperture perfect lens is sometimes included in the plot.

An alternative approach may be adopted in certain circumstances where it is required to show the form of the complex characteristics of the function. In this case the plot employs the conventional Argand diagram axes to represent the real and imaginary parts, taking the ordinate for the imaginary part $\mathscr{L}^i(u)$ and the abscissa for the real part $\mathscr{L}^r(u)$. The complex numbers representing the response for different frequencies are plotted and the resulting points joined to make a smooth curve. The curves are only of much value when the image suffers from a significant amount of an assymetrical aberration such as coma, since this leads to phase errors of some magnitude. Goodbody (1960), for example, employed the Argand diagram for presentation of the results in his study of the influence of coma on the response function. From these curves he was able to draw general conclusions about the quality of the optical image in the presence of coma. He deduced that in regions of frequency where the curve formed a loop one may expect poor imagery, but if the phase error, that is, the argument of the transfer function, is proportional to the spatial frequency, then the image merely suffers a bodily displacement without a relative shifting of the frequency component.

So far consideration has been given only to display of the OTF at one point in the image field and in one plane of focus. Generally, however, the variation with the pupil azimuth and the field angle is of considerable significance even in a corrected system. In order to simplify matters, the variation with pupil azimuth is examined only in the meridian and sagittal sections. Also, in a

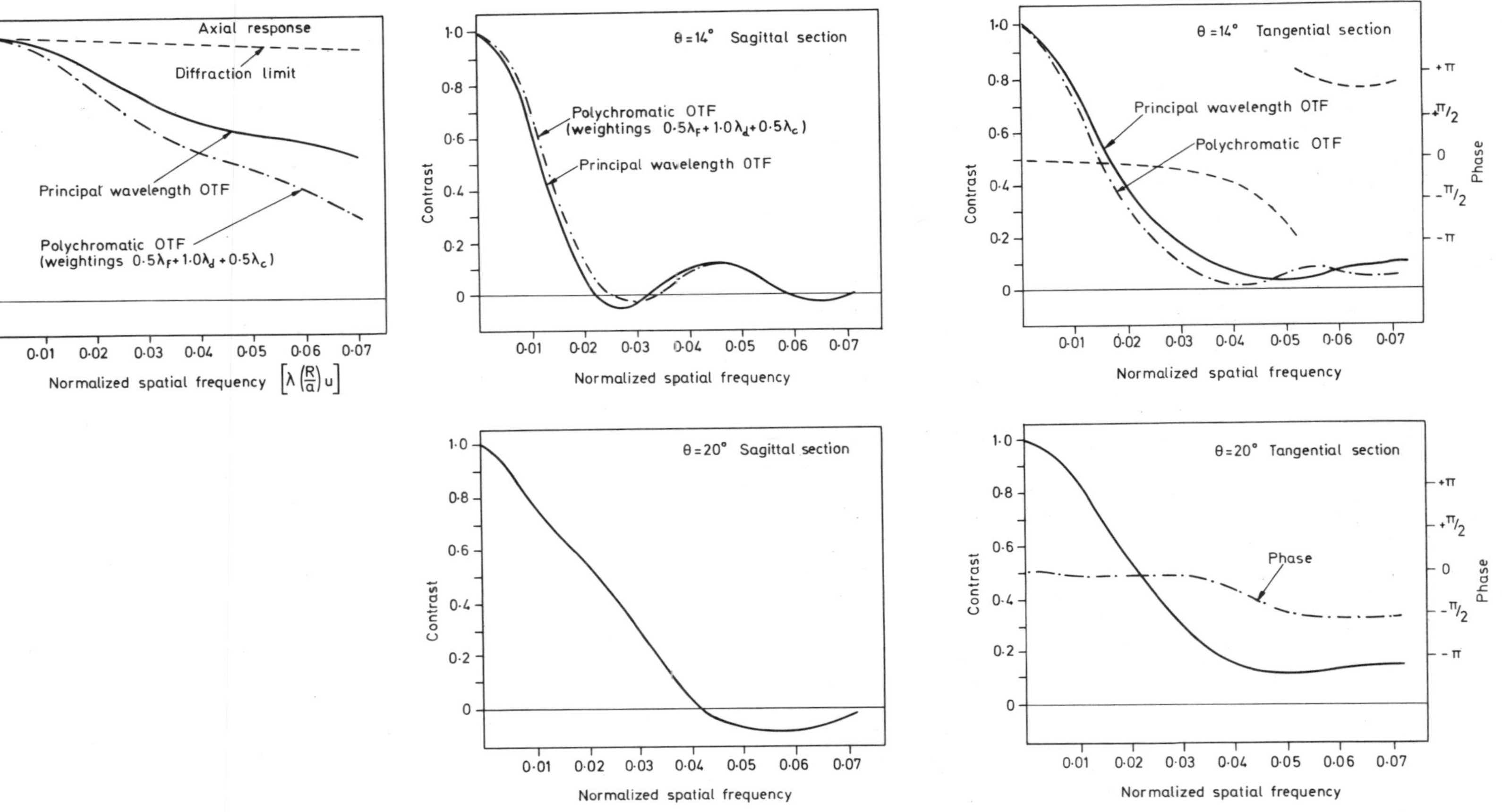

FIG. 24(*a*). Geometrical OTF in the plane of best focus (0·002 focal units inside the paraxial focus). Colour weightings are included for the polychromatic curve.

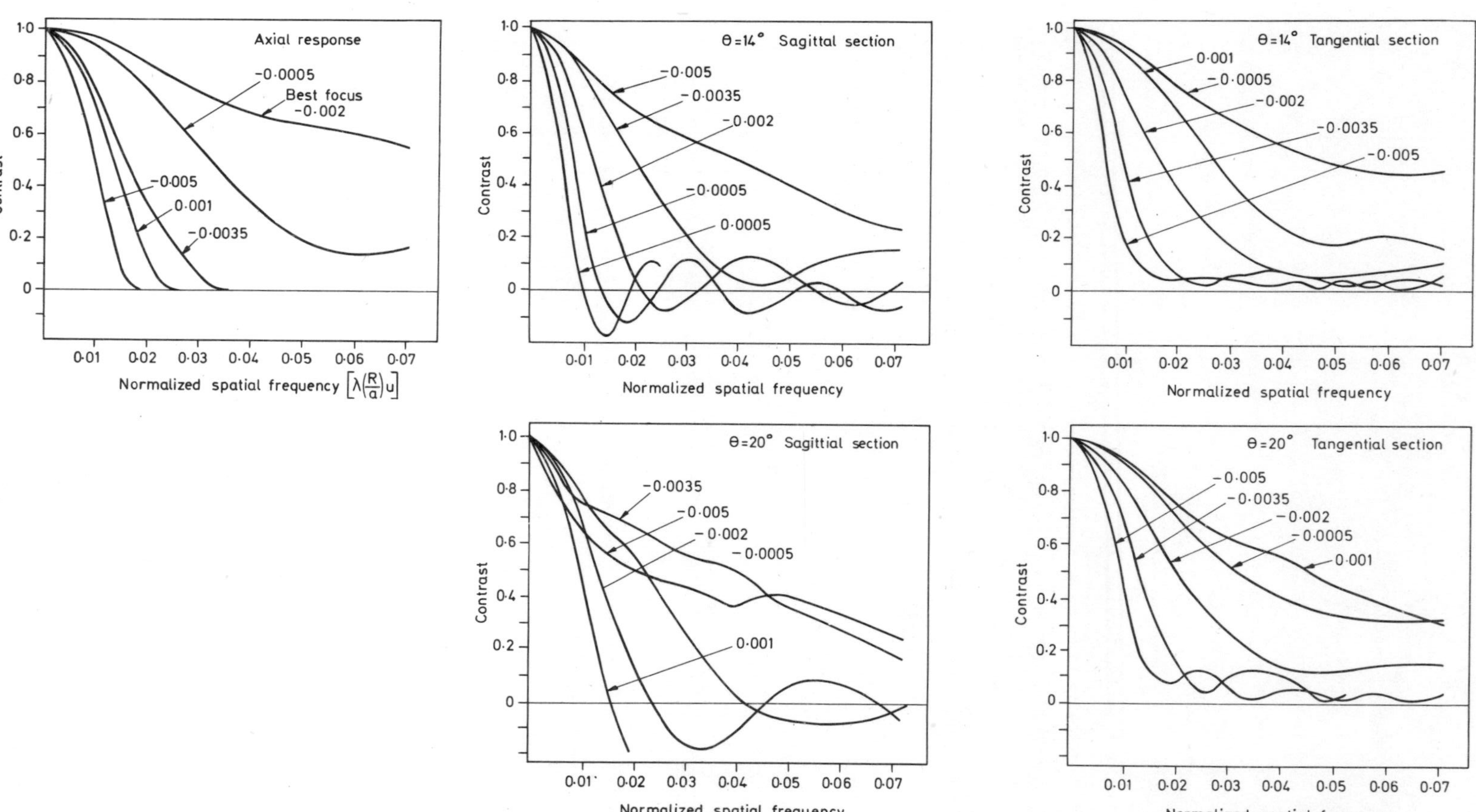

FIG. 24(*b*). Geometrical OTF showing the effect of change of focus. The ray tracing is limited to the principal wavelength, λ_d.

centred system extra-axial image points need be considered in the meridian section alone since all other orientations are represented by a rotation of the axes of the system. The choice of focal plane has a most important effect on the response function, so it is desirable to indicate graphically the form of the variation to assist in the selection of the plane of best focus. We begin to see that a whole series of curves are necessary in order to specify the system completely.

One means of greatly increasing the information contained in the graphical display is to introduce a third dimension into the plot. This may be achieved by plotting contrast solids for the various parameters. Steel (1956) has given a three-dimensional perspective drawing showing the effect of defect of focus on the axial response function. Fig. 25 is a reproduction of this drawing from which it will be seen that a plot of this kind requires considerable skill in draughtsmanship. The quantities included in the plot are modulation, frequency and defocus so that a family of solids are necessary to represent the off-axis performance.

Shannon has (1969) shown examples of perspective drawings

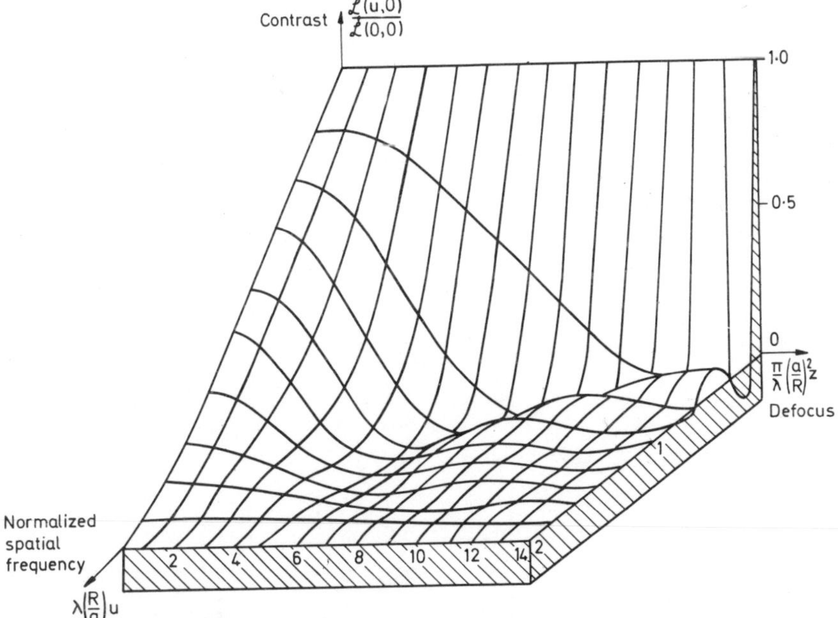

FIG. 25. An axial OTF solid for incoherent illumination of a system free of aberration but suffering from defect of focus, showing the normalized frequency response as a function of the spatial frequency *f* and of defocus *z*. (After Steel, 1956.)

to illustrate the possibilities of using an electronic computer to generate the plotting data. The computer then uses the data to drive a $X-Y$ plotter or cathode-ray display. The three-dimensional characteristics are clearly indicated in these beautiful pictures of contrast solids.

A much simpler solution to the problem of a presentation allowing for the extra parameters is available to those lacking the facilities of a large computer and expensive peripheries. The three-dimensional characteristics can be indicated by the cartographers trick of representing a solid in one plane by use of contours of height or, in this case, contrast. For example, using a rectangular co-ordinate system we may represent the field angle as the abscissa and defect of focus as the ordinate. The contrast loci are formed by joining points of equi-contrast calculated for a particular frequency. A set of contour maps are required to cover a range of frequencies and two pupil azimuths. In this way a vast amount of information can be presented in a very few graphs. A display of this kind was prepared for the double Gauss example using the results calculated by the geometrical approximation to the OTF. The contour maps shown in Fig. 26, therefore, correspond to the plots shown in Fig. 21. The technique used is simple but tedious by manual methods.

The basic spot diagram information is calculated by ray tracing and transformed to the geometrical OTF by the method described in § 3.4.5. The axis and, say, five field angles may be traced to provide the minimum data adequate for this purpose. At each field position the focal shift relations described in § 2.3.5 are used to generate the spot diagrams for each of, say, five planes of focus on each side of some chosen focal plane. A table of results can be drawn up giving the contrast at each of the field angles and planes of focus for a particular frequency chosen for the contrast map. To assist in the interpolation of the results a number of graphs should be drawn from the data in the table. These graphs are plots of contrast against field angle at each plane of focus. A similar set of graphs is drawn for the contrast against defocus at each field angle. Each of the graphs thus constructed represents sections through the contrast solid in directions perpendicular to the plane of the final contour map. The two forms of plot are also mutually perpendicular and provide the basic data for constructing the contour map. The interval of the contours must be chosen to give, say, contours at 1·0, 0·8, 0·6, 0·4 and so on. To find the locus of points at a particular contrast the graphs are examined to discover the co-ordinates in the plane of the contour map that correspond to this contrast. From the graphs representing the two mutually perpendicular sections scans can be made in directions parallel to the axes of the contour map, thus avoiding any problem of lack of information in regions where the scan is parallel to a contour of contrast. When all the points for a particular value of contrast have been plotted it is easy to draw a smooth curve through the points. This procedure is repeated for other frequencies and pupil azimuths.

3.4.7. *Optical transfer function based tolerance criteria*

In § 3.3 we saw how the Strehl intensity ratio may be developed to give maximum permissible tolerances for high performance systems. Hopkins (1957) has applied a similar method to the OTF. He used the minimum square value of the aberration function to give approximate values of the relative transfer function and thereby was able to extend the range of validity of diffraction-based tolerances. The technique is most suitable for tolerancing the less highly corrected systems where interest is mainly in good performance at the lower spatial frequencies. It is interesting to note that the OTF thus becomes complimentary to the Strehl intensity ratio in providing aberration tolerances, a point reinforced by Hopkins' suggestion that the Maréchal treatment of the Strehl tolerance is more appropriate to certain high performance systems such as high aperture microscope objectives, because of approximations made in evaluating averaging integrals. It may be desirable in many practical systems not needing the highest form of aberration correction to ensure good contrast at low spatial frequencies. This is necessary to prevent degradation by the optical system of the overall performance which includes the transfer function of both the optical system and the detector. For example, the MTF of a photographic emulsion tends to drop rapidly at low spatial frequencies and then level off over the middle range of spatial frequencies. Optimum performance is obtained by designing a

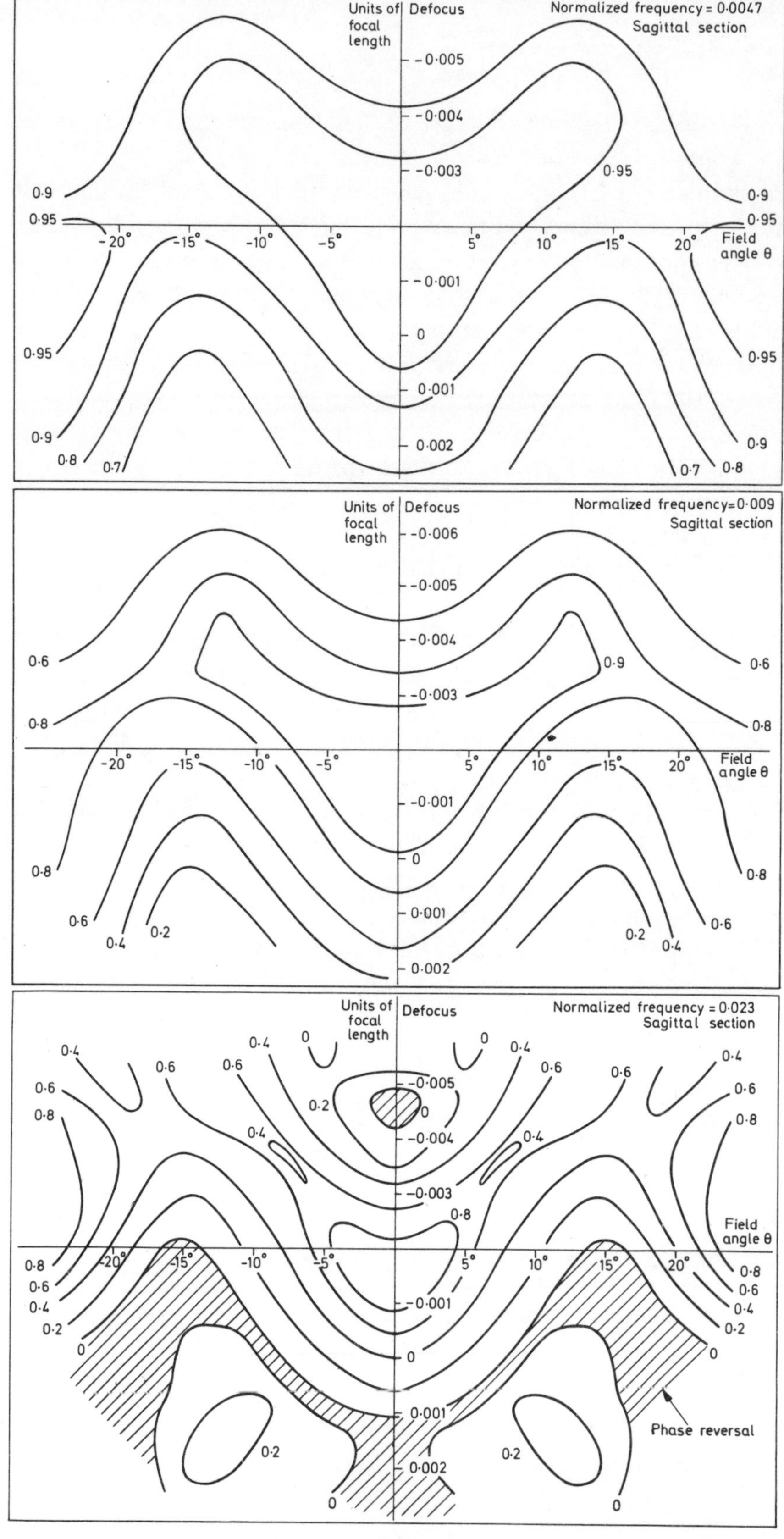

FIG. 26. OTF solids drawn as contours of contrast, based on the geometrical OTF of the Wynne double

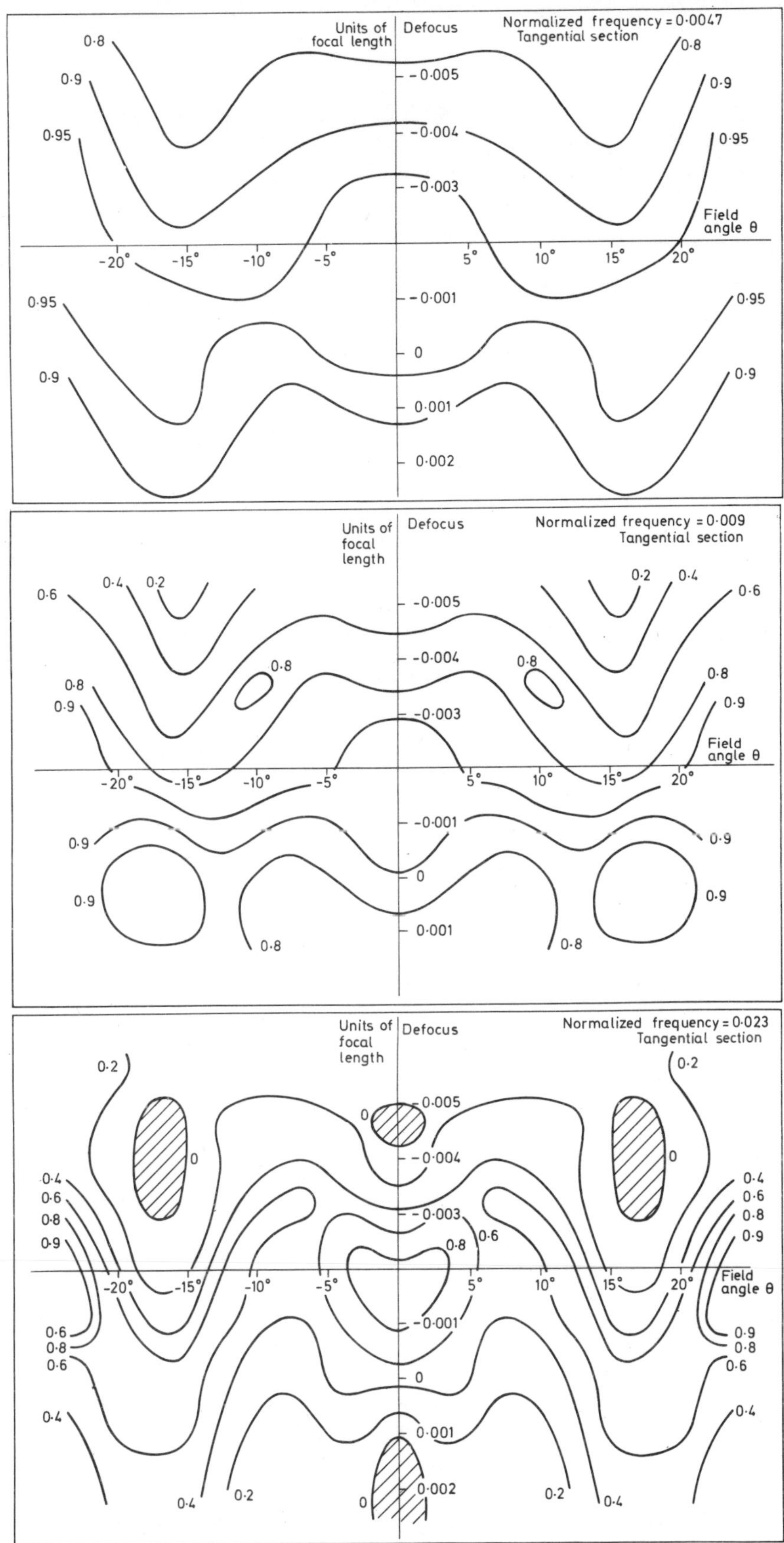

Gauss design. Here the frequency response is shown as a function of field angle and defect of focus.

lens with nearly unit contrast at low spatial frequencies, thereby conserving the response of the photographic emulsion. Therefore, some significance can be attached to simple tolerance formulae which determine low frequency performance. Hopkins has provided suitable tolerances in his approximate formulae developed for spatial frequencies up to $s = 0{\cdot}20$.

Hopkins defines the mean value of the aberration function to be

$$\overline{W(s, \psi)} = \iint_S W(x, y; s)\, dS$$

where s is a normalized line frequency obtained from the spatial frequencies (u_0, v_0) by a rotation ψ of the system axes. The effective region of integration for the reduced pupil co-ordinates (x, y) is defined by the common area of the sheared pupil and denoted by S, with the pupil element dS being defined by $dS = (dx\, dy)/S$. Next, Hopkins defined the relative response $M(s, \psi) \exp [i\theta(s, \psi)]$ to be the ratio of the response in the presence of aberration and defect of focus, to the response obtained for the same system without aberration and having a focal plane at the diffraction focus. Hopkins' tolerances, therefore, employ a relative frequency response in place of the ratio of intensity adopted by Maréchal to derive the Strehl tolerances. The approximate equation given by Hopkins may be written as

$$M(s, \psi) \exp [i\theta(s, \psi)] = \exp [iks\overline{W(s, \psi)}] \left[1 - \frac{2\pi^2 s^2}{\lambda^2} K(s, \psi) \right] \quad (3{\cdot}5)$$

where the minimum mean square value of $W(x, y; s)$ is

$$K(s, \psi) = \iint_S W^2\, dS - \left(\iint_S W\, dS \right)^2 \quad (3{\cdot}6)$$

The relative contrast modulation $M(s, \psi)$ and the phase shift $\theta(s, \psi)$ may be separated by virtue of the complex character of equation (3.5);

$$M(s, \psi) = 1 - \frac{2\pi^2 s^2}{\lambda^2} K(s, \psi) \quad (3{\cdot}7)$$

and

$$\theta(s, \psi) = ks\overline{W(s, \psi)} \quad (3{\cdot}8)$$

Hopkins showed that (3.7) may be expressed as an inequality

$$M(s, \psi) \leqq 1 - \frac{2\pi^2 s^2}{\lambda^2} K(s, \psi)$$

provided that the right-hand side is greater than zero. He then proposed the condition $M(s, \psi) \geqq 0{\cdot}80$ as a criterion for permissible aberration in an optical system. Hence the tolerance condition for the single frequency s with orientation ψ is

$$K(s, \psi) \leqq \frac{0{\cdot}10\lambda^2}{\pi^2 s^2}$$

When this equation is satisfied, the contrast of the image of the component of spatial frequency s is not less than 0·80 times the contrast that would have been obtained for that frequency at the diffraction focal plane and in the absence of aberration.

Calculating $K(s, \psi)$ and hence $W(s, \psi)$ requires evaluating integrals of the form

$$H_{kl}(s) = \iint_S x^k y^l \, dS$$

Hopkins has done this for defocus, spherical aberration, coma, astigmatism and field curvature. He presented the results graphically for the complete range of spatial frequencies, showing the appropriate balance of aberration for such mixtures of aberration as defocus, and primary and secondary spherical aberration. The results are summarized in the set of tolerances listed below, and are given in the form of approximate expressions of sufficient accuracy for most purposes, since the error is less than $0{\cdot}10\lambda$ provided $s < 0{\cdot}20$.

1. *Tolerance for defect of focus.* In the absence of any other deformation of the wavefront, the aberration function in the presence of focus is of the form

$$W(x, y) = A_{20}(x^2 + y^2)$$

and the tolerance formula for the aberration coefficient A_{20} becomes

$$A_{20} = \pm \frac{0{\cdot}10}{s} \lambda$$

Using the substitutions

$$A_{20} = \tfrac{1}{2} n' \sin \alpha \, \delta\zeta' \quad \text{and} \quad s = \frac{\lambda}{n \sin \alpha} g = \frac{\lambda}{n' \sin \alpha'} g'$$

where $\delta\zeta'$ is the axial displacement, g and g' are the number of cycles per unit length of sinusoidal line intensity distributions in the object and image spaces respectively, and $n \sin \alpha$ and $n' \sin \alpha'$ are the corresponding apertures, we can replace the aberration coefficient A_{20} by the axial displacement

$$\delta\zeta' = \pm \frac{0 \cdot 20}{g' \sin \alpha'}$$

2. *Tolerances for spherical aberration.* Since there is no field dependence, the aberration function in the presence of various orders of spherical aberration may be written in the two-term polynomial

$$W(x, y) = A_{20}(x^2 + y^2) + A_{40}(x^2 + y^2)^2 + A_{60}(x^2 + y^2)^3 + \ldots$$

including the defocus term. The tolerance for secondary spherical aberration becomes

$$A_{60} = \pm \left(\frac{0 \cdot 58}{s} + 1 \cdot 5 \right) \lambda$$

with the balance of aberrations in the ratios

$$\frac{A_{40}}{A_{60}} = -(1 \cdot 80 - 1 \cdot 5s + 1 \cdot 5s^2)$$

$$\frac{A_{20}}{A_{60}} = +(0 \cdot 90 - 1 \cdot 4s + 2 \cdot 1s^2)$$

When $A_{60} = 0$, the tolerance on spherical aberration is

$$A_{40} = \pm \left(\frac{0 \cdot 212}{s} + 0 \cdot 33 \right) \lambda$$

with the defocus term only

$$\frac{A_{20}}{A_{60}} = -(1 \cdot 33 - 1 \cdot 1s + 0 \cdot 7s^2)$$

3. *Tolerance for coma.* The general form of the aberration function was given in § 3.3, but, for present purposes, we may express the circular coma in a two-term polynomial. Neglecting the field dependence we have

$$W_0(x_0, y_0) = A_{31}(x_0^2 + y_0^2)y_0 + A_{51}(x_0^2 + y_0^2)^2 y_0$$

for the aberration referred to the paraxial image point. To allow for the use of line frequencies, Hopkins replaced this expression by

$$W(x, y) = A_{31}(x^2 + y^2)(y \cos \psi + x \sin \psi) + A_{51}(x^2 + y^2)^2(y \cos \psi + x \sin \psi)$$

In the presence of coma the tolerance formulae depend on whether the line structure is lying parallel with the axis of symmetry of the comatic image ($\psi = 0$) or in the sagittal direction ($\psi = \frac{1}{2}\pi$). In general, for the latter case, there will be a phase shift $\theta(s, \psi)$. Therefore, the two forms of tolerances are:

For the azimuth $\psi = 0$,

$$A_{51} = \pm\left(\frac{0{\cdot}64}{s} + 0{\cdot}96\right)\lambda$$

With the balance of aberrations in the ratio

$$\frac{A_{31}}{A_{51}} = -(1{\cdot}49 - 0{\cdot}94s + 0{\cdot}41s^2)$$

When $A_{51} = 0$, the primary coma is given the tolerance

$$A_{31} = \pm\left(\frac{0{\cdot}246}{s} + 0{\cdot}19\right)\lambda$$

For the azimuth $\psi = \frac{1}{2}\pi$, the corresponding values are

$$A_{51} = \pm\left(\frac{0{\cdot}37}{s} + 0{\cdot}83\right)\lambda$$

the phase shift being $\theta(s) = \pm(1{\cdot}16 + 0{\cdot}35s)$, with the balance of aberrations in the ratio

$$\frac{A_{31}}{A_{51}} = -(1{\cdot}49 - 1{\cdot}26s + 0{\cdot}85s^2)$$

When $A_{51} = 0$, the primary coma is given the tolerance

$$A_{31} = \pm\left(\frac{0{\cdot}142}{s} + 0{\cdot}16\right)\lambda$$

the phase shift being $\theta(s) = \mp(0{\cdot}89 + 0{\cdot}24s)$

4. *Tolerance for astigmatism and field curvature.* The terms in the wave aberration function representing astigmatism and field curvature can be expressed thus:

$$W_0(x_0, y_0) = A_{20}(x_0^2 + y_0^2) + A_{22}y_0$$

which again, according to Hopkins, can be transformed by the rotation ψ giving

$$W(x, y) = A_{20}(x^2 + y^2) + A_{22}(x^2 \sin^2 \psi + xy \sin 2\psi + y^2 \cos^2 \psi)$$

The aberration coefficients A_{20} and $A_{20} + A_{22}$ measure the difference in focus between the chosen image plane and the sagittal and tangential foci respectively. If the line frequencies are chosen to be parallel with either the sagittal or tangential focal line, the tolerance for astigmatism is the same as for the simple case of defect of focus

$$A_{20} = \pm \frac{0{\cdot}10}{s} \lambda$$

and the image will be free of phase shift. At the appropriate focal line, of course, the image of line structures will be unaffected by astigmatism. Hopkins also showed that the relative contrast modulation averaged over all azimuths, $\overline{M(s, \psi)}$, has a maximum value when $A_{20} = -\frac{1}{2}A_{22}$, a result for best focus which agrees with the one predicted by the Strehl calculation. This condition implies the best focal plane lies midway between the two focal lines. At this focal plane the approximate tolerance formula becomes

$$A_{22} = \pm \frac{0{\cdot}20}{s} \lambda$$

with the separation between foci being

$$\delta\zeta_A = \pm \frac{0{\cdot}40}{g' \sin \alpha'}$$

for low spatial frequencies ($s \leqq 0{\cdot}10$) and within 10 per cent error for spatial frequencies up to $s = 0{\cdot}20$.

In many simple lens types such as doublets and Petzval lenses the aberration coefficients are related thus:

$$A_{20} = A_P + \tfrac{1}{2}A_{22}$$

where A_P is the coefficient of primary field curvature, measured by the shift of focus from the paraxial focal plane to the Petzval surface. The best one can achieve in these designs is the optimum balance of the primary astigmatism and field curvature. Hopkins' treatment suggests that, for a paraxial focus and small values of s, the relationship becomes

$$A_{22} = -\tfrac{4}{5}A_P$$

which is independent of frequency. Substituting back into the above equation we obtain

$$A_{20} = -\tfrac{3}{5}A_P \quad \text{and} \quad A_{20} + A_{22} = -\tfrac{1}{5}A_P$$

These expressions indicate the two foci for the sagittal and tangential sections are on either side of the paraxial focal plane, separated in the ratio 3 : 1. In these circumstances Hopkins' tolerance for field curvature is

$$A_P = \pm\frac{0{\cdot}22}{s}\lambda$$

which corresponds to a longitudinal focal error $\delta F'$ given by

$$\delta F' = \pm\frac{0{\cdot}44}{g' \sin\alpha'}$$

Therefore, the tolerance for field curvature is double that for defect of focus, provided the best state of correction for astigmatism relative to the paraxial focal plane has been adopted.

A combined tolerance for defect of focus and Petzval curvature is obtained by summing the two tolerances

$$\delta\zeta' + \delta F' = \pm\left(\frac{0{\cdot}20\lambda}{g' \sin\alpha'} + \frac{0{\cdot}44\lambda}{g' \sin\alpha'}\right) = \pm\frac{0{\cdot}64\lambda}{g' \sin\alpha'}$$

Note that the coefficient of field curvature A_P now includes an extra amount of focal defect, so the relationship $A = -\tfrac{4}{5}A_P$ must change, leading to a different astigmatism balance.

In practice, most optical systems are required to image a range of spatial frequencies, so it is of interest to know how these tolerances should be applied to actual systems in view of the fact that the tolerances have been given as functions of spatial frequency. Hopkins' approximate formulae indicate that in the range $0 < |s| < 0{\cdot}20$ the permissible aberration decreases with increasing frequency. The value of s to be substituted in the formulae is, therefore, the maximum spatial frequency the system is required to transmit. For the region where the value of s is greater than 0·20, Hopkins' graphs show the tolerances tend to be constant up to about $s = 1{\cdot}5$ and then to increase rapidly with frequency. Although these tolerances can be shown to be substantially equal to those obtained by Maréchal, the latter are in fact, for computational reasons, more suitable for the high frequency domain.

3·5. ENCIRCLED ENERGY (TOTAL ILLUMINATION)

The determination of the light intensity distribution in a star-image is essentially one of evaluating a double integral over the pupil plane. This has been carried out numerically by several authors (Barakat, 1961a, gives a bibliography) for various special cases. It is usual to plot the intensity distribution as contours of equal intensity (isophots), many examples of which are reproduced by Linfoot (1955).

The intensity distribution of a star-image produced by an aberration-free system having a circular pupil is, of course, the Airy disk. This can be represented by the Bessel function of the first kind $J_1(z)$. Let the pupil radius be a, the radius of the reference sphere be R, and the medium of the image space have refractive index n' at wavelength λ. Then the intensity distribution, normalized to unity at the centre, is given by

$$I(z) = \left[\frac{2J_1(z)}{z}\right]^2$$

where

$$z = \frac{2\pi n' a}{\lambda R}$$

is the generalised z unit which is characteristically independent of the scale of the system.

This is an undulating function consisting of a central core surrounded by a series of bright rings which rapidly decay to insignificantly low intensity. As aberration is introduced into the system, energy spreads into the outer regions. Lord Rayleigh (1881) first suggested that it would be of interest to know the proportion of the total energy in the image contained within any one ring. This quantity is now known as the encircled energy (or by less informative term 'total illumination'). Thus, the encircled energy describes the average integrated behaviour of the point source diffraction image as modified by the aberration. This is a smooth function in comparison with the point source diffraction image and is therefore more easily evaluated by numerical methods. The encircled energy in a ring of radius z_0 is defined in polar co-ordinates by the equation

$$E(z, W)=N\int_0^{z_0} I(z, W)z\, dz$$

where the intensity point spread function

$$I(z, W)=\left|\int_0^1 J_0(z, \rho)\rho \exp\,[ikW(\rho)]\, d\rho\,\right|^2$$

N is a normalizing constant, and $W(z, \rho)$ the aberration function. In Fig. 27(*a*) and (*b*) the form of the encircled energy is shown for the integrated Airy disk, the ideal case. The other curves indicate the form of the encircled energy when the aberration function $W(\rho)$ has a rotationally symmetric form

$$W(\rho)=\sum_2^N W_{2n}\,\rho^{2n}$$

Fig. 27(*c*) indicates the form of the curves taken when the effect of longitudinal focus is mixed with primary spherical aberration; $\beta_{40}=\frac{1}{2}$; $\beta_{40}=1$ and $\beta_{40}=2$. The variable β is the coefficient, measured in units of wavelength, of the power series representing the wavefront aberration. The curves showing defocus are due to Wolf (1951), and those showing spherical aberration are due to Barakat (1961a). In his paper, Barakat describes the numerical evaluation of the encircled energy using seven point Gauss quadrature. He used forty-nine quadrature points of which only twenty-eight were calculated owing to the system's symmetry. The spherical aberration was represented by Zernike polynomials. A later paper by Barakat and Movello (1964) describes the evaluation of encircled energy via diffraction theory from the system data of a number of designs. The systems were ray-traced, the wavefront polynomial obtained and from this, the encircled energy, OTF and intensity distribution were calculated.

Again, we find, as with OTF, it is much simpler and hence more economical on computer time to use the geometrical

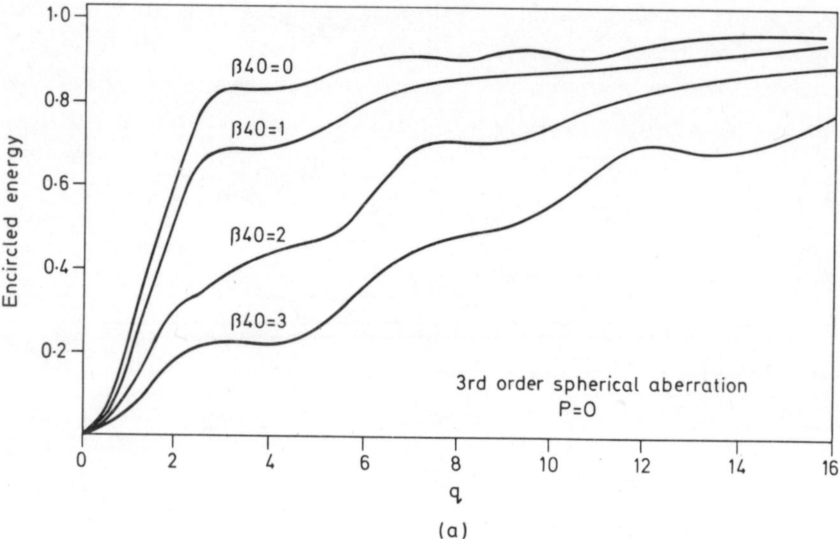

FIG. 27(*a*). Encircled energy (total illumination) for various amounts of primary spherical aberration as indicated by the coefficient of the wavefront aberration β_{40}. The energy is calculated for the diffraction focus.

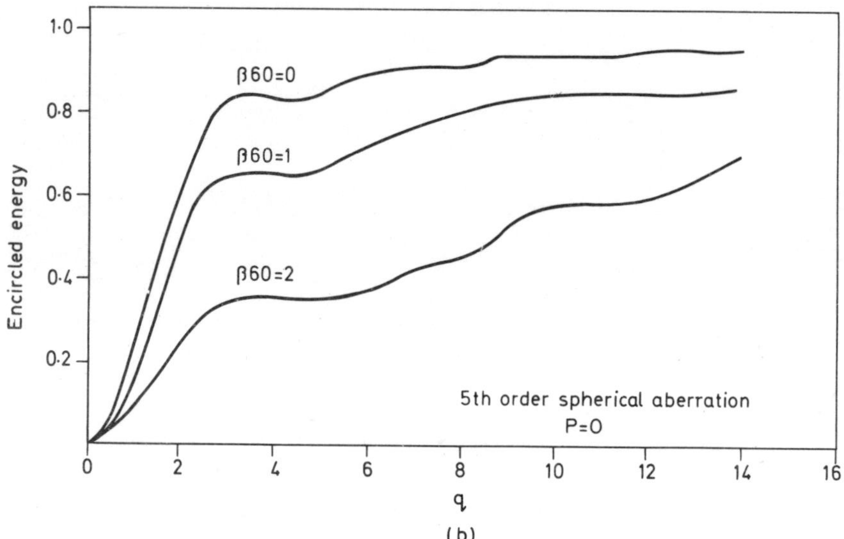

FIG. 27(*b*). Encircled energy for various amounts of secondary spherical aberration as indicated by the magnitude of the coefficient of the wavefront aberration β_{60}. The energy is calculated for the diffraction focus.

FIG. 27(*c*). (p. 93) Encircled energy as a function of defect of focus. The contours of encircled energy illustrate the increasing asymmetry with the increase in the magnitude of the spherical aberration. The points P and M are the paraxial and marginal foci respectively.

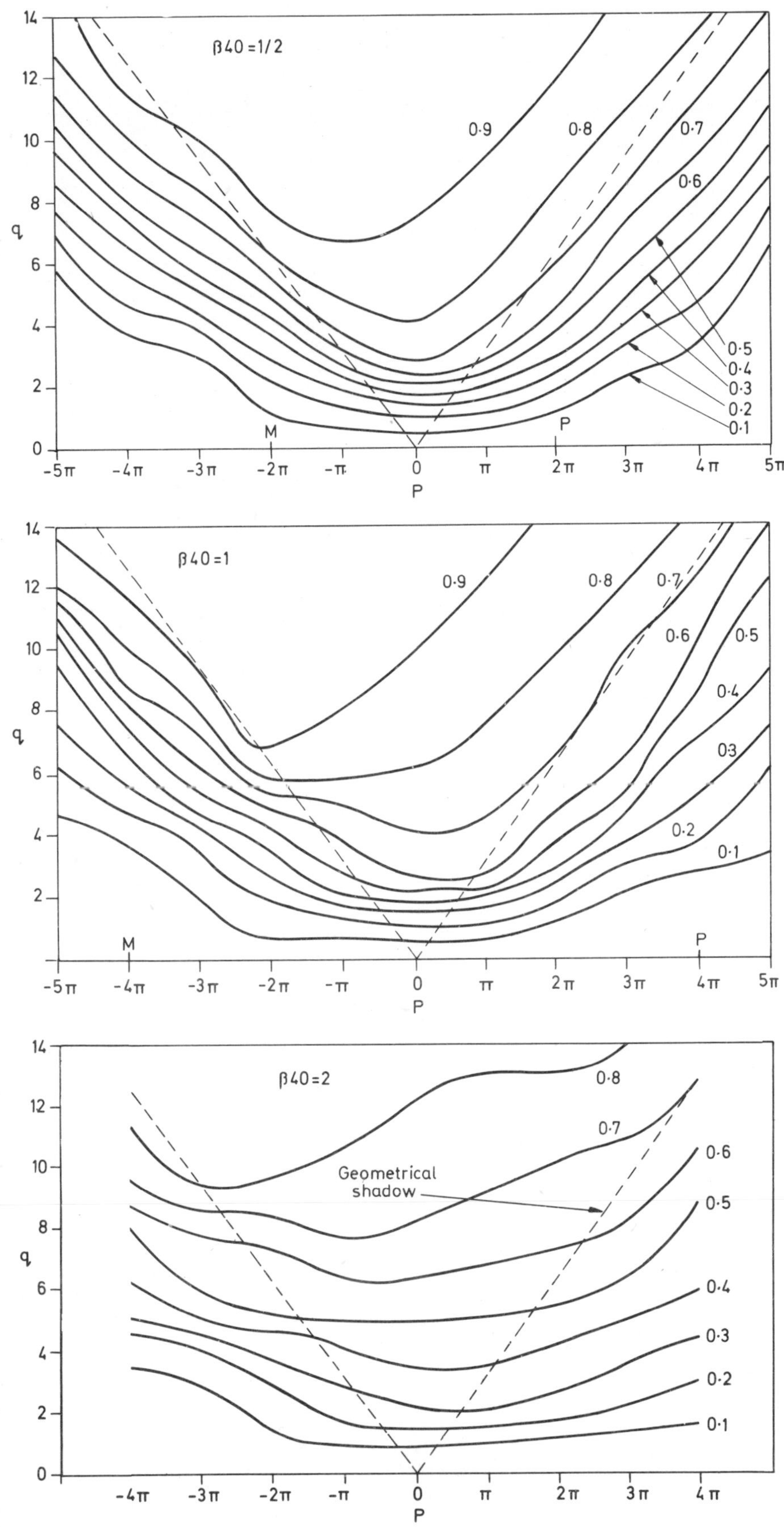
β40 = 1/2
β40 = 1
β40 = 2
q
P
M
Geometrical shadow
0·1
0·2
0·3
0·4
0·5
0·6
0·7
0·8
0·9
-5π
-4π
-3π
-2π
-π
0
π
2π
3π
4π
5π

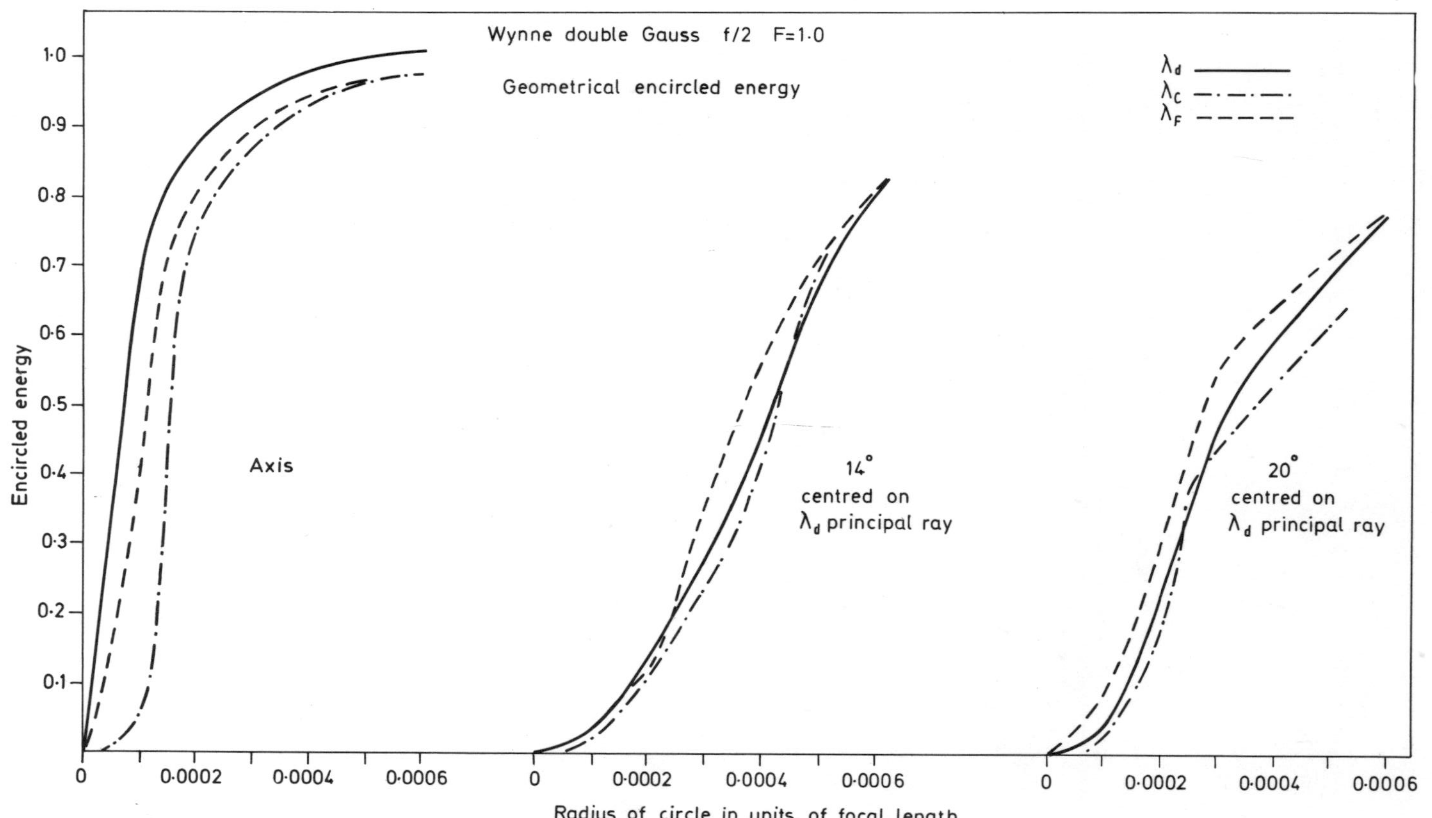

FIG. 28. Geometrical approximation to encircled energy calculated for the Wynne double Gauss example. Analysis is at the paraxial focal plane. The extra-axial analysis is arbitrarily centred about the principal ray interception in the paraxial focal plane.

approximation to the intensity distribution of a point image described in § 2.3.4 in connection with transverse aberration polynomials and spot diagrams. In this application the density of spots within a given radius of the image point is determined and the spot density plotted against the radial distance. The geometrical approximation to encircled energy is valid only for systems having relatively large residual aberration, as in the case of the double Gauss example considered previously. In order to demonstrate the form such a plot might take in typical case, a computer program was written and used to evaluate the encircled energy axially and at two field points for the double Gauss. These are shown in Fig. 28.

It is possible to derive a tolerance from the encircled energy method of representation. R. E. Hopkins (1962) has used an optimization figure of merit based on the radius of the circle obtaining 30 per cent of the total energy. Other similar criteria may be developed for particular design problems.

4

Interferograms

Interferometry is widely recognized as an effective technique for testing optical systems. This acknowledges the fact that the asphericity of the image-forming wavefront is the significant quantity in determining the performance of an optical system. There are a variety of wavefront-testing interferometers used for this purpose. Discussion of interferometry is out of place here, but has been fully described by many authors, such as Steel (1967) and Françon (1966).

An interferogram is, in essence, a one-dimensional representation of the shape of the wavefront in the lens pupil. An interference fringe maps the locus of points of equi-phase determined by the amplitude summation of the reference sphere and the wavefront under investigation. It is generally most convenient to examine fringes in star-space where the reference wavefront becomes a plane wave. Under these circumstances the interpretation of the interferogram is simplest; the fringes follow the loci of the points of equi-phase, or phase contours produced by the intersection of the image-forming wavefront and a series of parallel planes separated by an interval equal to some submultiple of the wavelength of light employed.

Frequently we know the constructional data of an objective and the aim of interferometric testing is to determine the effect of manufacturing faults in the performance of the system. In such cases we require to know the shape of the wavefront in the exit pupil of the correctly centred objective, that is, the objective with perfect spherical (or controlled aspheric) surfaces, homogeneous glass and components of correct thickness, correctly set in position.

From the shape of the wavefront of this hypothetical objective, we can compute the theoretical interferogram. Any manufacturing defects resulting in departure of the manufactured lens from the hypothetical one will cause the interferometer to produce a slightly different interferogram. It may then be possible to set a tolerance on the variation from the theoretical interferogram or even to investigate the sources of the variations by making changes to the input data to the computations. It is comparatively easy to compute the interferogram for the axial case with no tilt to the intersecting plane, as the system has rotational symmetry and so the interferogram consists of concentric circles. The radii of the circles can be obtained from a graph of the wavefront aberration calculated in the meridian section. However, as soon as tilt of the intersecting plane or extra-axial object points are considered, the system loses symmetry, and the calculations become tedious, if not impossible, by manual methods. It is obviously a problem for an electronic computer.

A computer program for generating interferograms has been written in the Fortran Language for use on an IBM 7094 machine, and this chapter is a description of this program and its use.

4.1. THE PROBLEM

The usefulness of an interferometric test is greatly extended if the shape of the interferogram pattern is known for the objective free from manufacturing faults. Therefore, the task to be given to the computer is the generation of such a pattern from the design data of the objective, that is, the curvatures, thicknesses, separations and refractive indices.

The interferogram may be produced by an $X - Y$ graph plotter or on a line printer employing printing characters to represent the fringe contours. The latter course was chosen because line-printers are more likely to be available. In either case, the system must be ray-traced and the wavefront aberration determined. The shape of the wavefront in the exit pupil (ignoring pupil aberration) is then indicated by contours with an interval equal to one wavelength or by suitable scaling. The interval can be made to match the particular interferometer.

The calculation of the wavefront aberration may employ either the optical pathlength differences along rays or the method due to H. H. Hopkins. Also, the shapes of the wavefront may be determined by either tracing a small number of rays and fitting a polynomial, or tracing a larger number of rays in a pupil filling network. The actual choice of method is largely dependent on the accuracy required in the interferogram. The differencing of optical pathlength is subject to rounding errors. Hopkins' method probably doubles the ray tracing time and tracing a fine mesh of rays through the pupil for high accuracy is only accomplished by generating a very large number of rays which, in any case, is an inefficient course. Taking these factors into account, for example where the $X-Y$ plotter can be employed, it might be advisable to trace a moderate number of rays, fit a polynomial and hence interpolate at close intervals to drive the plotter. This would undoubtedly produce a very accurate interferogram but would probably tend to be wasteful of expensive computer time in view of the limited accuracy of, say, a twentieth of a wavelength obtainable on a practical interferometer.

4.2. THE PROGRAM

Basically, the program consists of a Fortran IV version of the method described in Chapter 2. The program is written in four parts.

1. The main program containing the ray generation method described in § 2.2.2, and the analysis of the wavefront aberration in terms of fringe contours.
2. A subroutine for paraxial ray tracing described in § 2.1.
3. A subroutine for skew ray tracing as described in § 2.1.
4. A subroutine for calculating the wavefront aberration from skew data as described in § 2.5.1.

A flow chart (Fig. 29) has been included in preference to a compilation because the actual programming is straightforward though less easy to follow. The program uses most of the 32 000 word store of the IBM 7094, since a number of arrays containing 57×95 elements are employed. A typical run time of 3 minutes produces interferograms of the double Gauss example for the axis and at two field angles with four tilt angles per image point. This is considered satisfactory, although the program does contain a number of inefficiencies which should be eliminated with further development. For example, one worthwhile gain already included was a routine to trace only half the pupil, thus taking advantage of symmetry about the meridian section in order to fill in the aberration of the other half of the pupil.

The input data are conventional. The heights and convergence angles of the paraxial principal ray and the paraxial image ray are specified at the object plane or first surface for an infinite object conjugate. The wavelength of light is specified in the same units as the system data. The number of field angles and number of angles of tilt required are specified, the latter being a displacement of the reference plane at the edge of the pupil given in units of wavelength. For the system data, there is a card for each surface giving the surface number, the curvature, the separation from the preceeding surface, the refractive index and the surface diameter for the vignetting test. In addition, a number of controls are included, such as optional printing of the area around the pupil and an optional square root density of field points. As the ray generation and ray tracing have been described in the Chapter 2, they will not be considered any further here. The only image space data stored by the program are the wavefront aberration and a label indicating whether or not the ray was transmitted. Both these quantities are stored in 57×95 element arrays.

The analysis of the interferogram is performed twice; the wave aberration array is first scanned row by row and then column by column. The reason for this becomes clear when the technique of analysis is understood. The scanning process involves a number of tests applied to each element of the array in turn. The first test is for the inclusion of the point within the circular pupil and if this holds a test is made for transmission through the system. Appropriate characters are stored in the printing array if these tests fail. Otherwise, the wave aberration in adjacent elements is compared and if the integer values are equal, a blank is stored, if not, the nearest integer value of their mean is calculated. Tests are made for positive and negative mean values; if positive, the appropriate integer is stored, or if negative, the integers are converted to alphabetic characters so that $A \equiv -1$, $B \equiv -2$, $C \equiv -3$, and so on. For a mean in excess of nine wavelengths, a plus sign is stored, while for negative

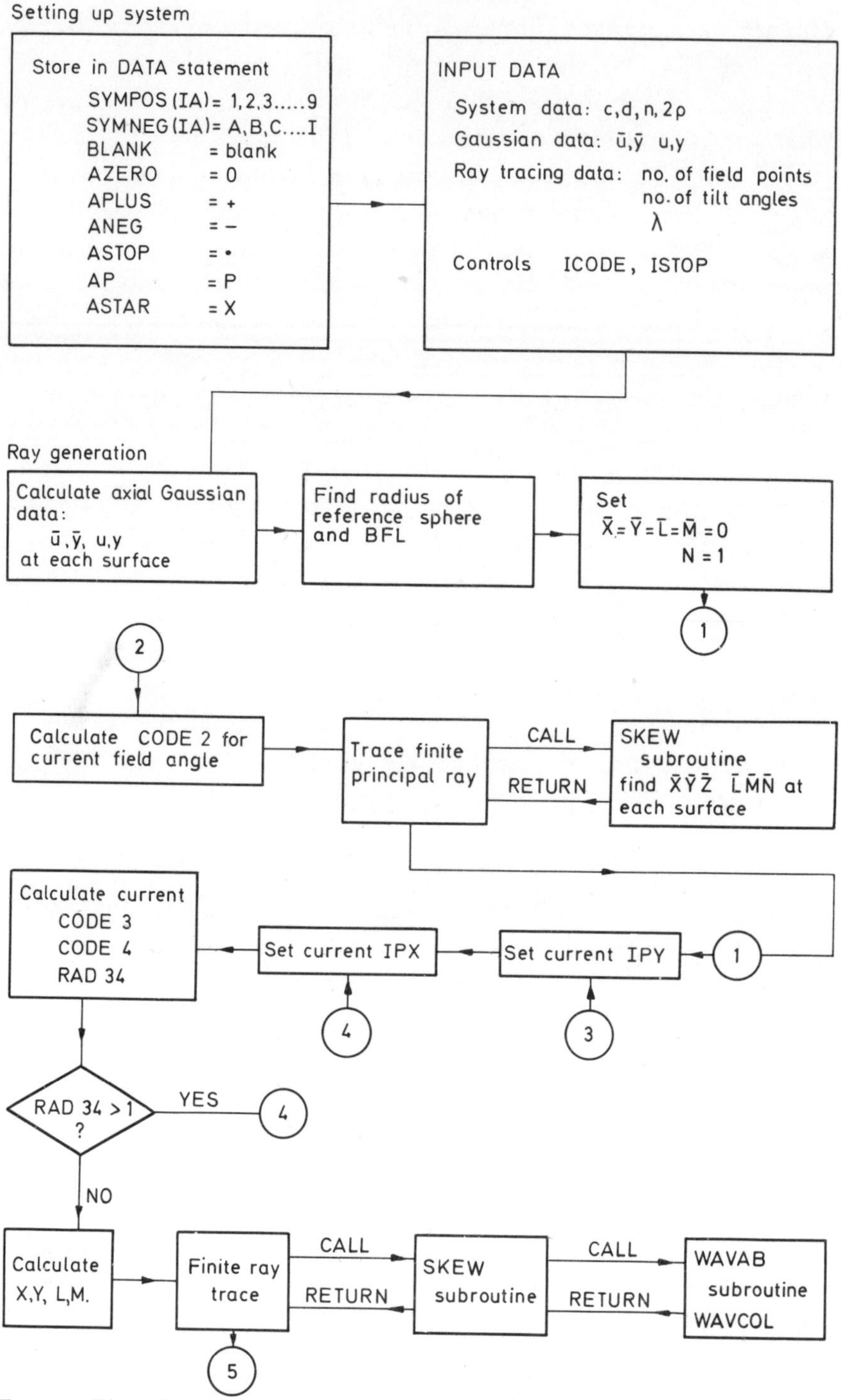

Setting up system

Store in DATA statement
SYMPOS (IA) = 1, 2, 3.....9
SYMNEG (IA) = A, B, C....I
BLANK = blank
AZERO = 0
APLUS = +
ANEG = –
ASTOP = •
AP = P
ASTAR = X

INPUT DATA
System data: c, d, n, 2ρ
Gaussian data: $\bar{u}, \bar{y}$ u, y
Ray tracing data: no. of field points
no. of tilt angles
λ
Controls ICODE, ISTOP

Ray generation

Calculate axial Gaussian data: $\bar{u}, \bar{y}$, u, y at each surface

Find radius of reference sphere and BFL

Set $\bar{X} = \bar{Y} = \bar{L} = \bar{M} = 0$
N = 1

(1)

(2)

Calculate CODE 2 for current field angle

Trace finite principal ray

CALL

RETURN

SKEW subroutine find $\bar{X}\bar{Y}\bar{Z}$ $\bar{L}\bar{M}\bar{N}$ at each surface

(1)

Set current IPY

(3)

Set current IPX

(4)

Calculate current
CODE 3
CODE 4
RAD 34

RAD 34 > 1 ?

YES

(4)

NO

Calculate X, Y, L, M.

Finite ray trace

(5)

CALL

RETURN

SKEW subroutine

CALL

RETURN

WAVAB subroutine WAVCOL

FIG. 29. Flow chart for the interferogram program employing the conventions of the Fortran language. The chart continues on pp. 101, 102, and 103.

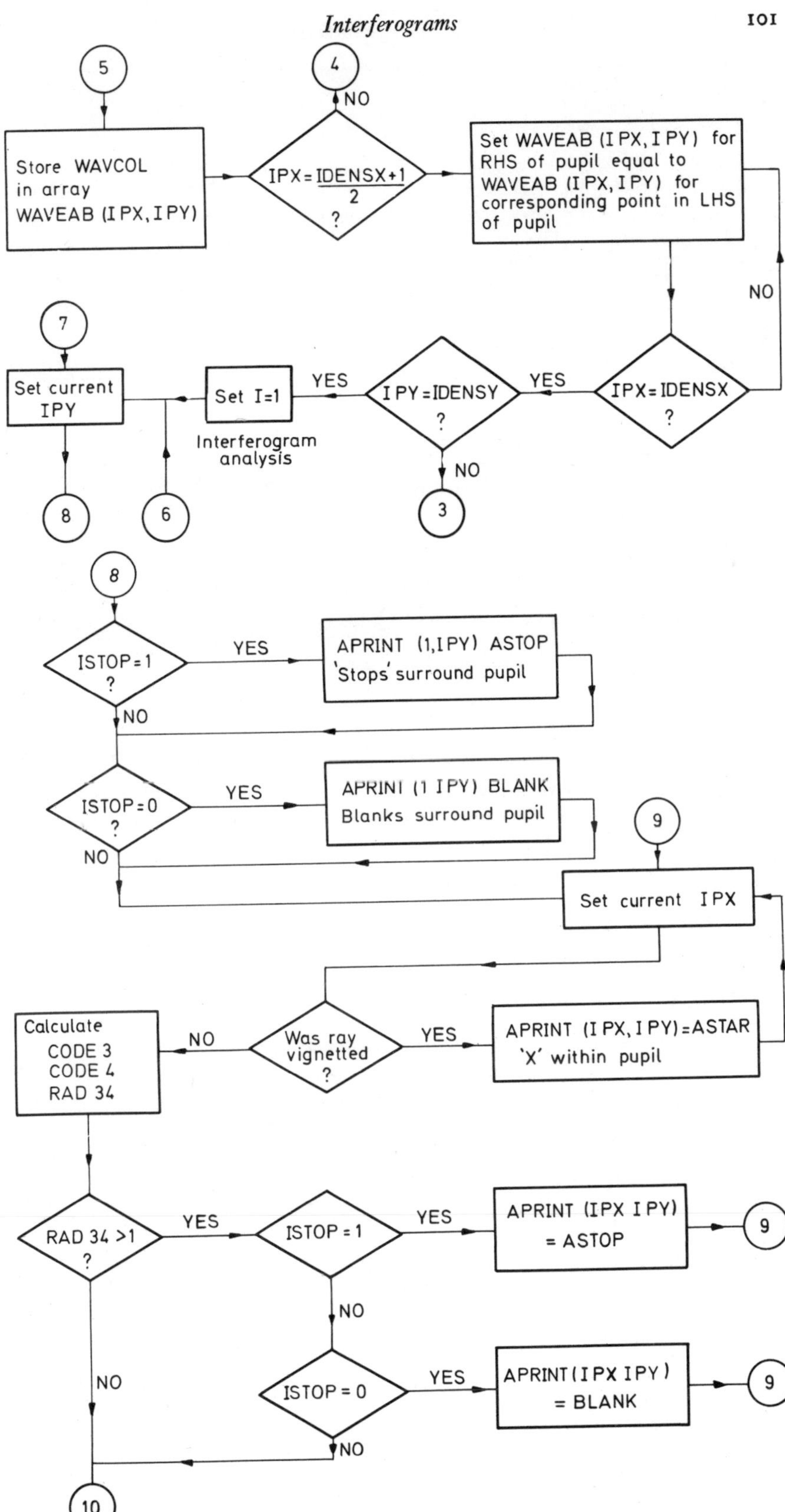
5
4
NO
Store WAVCOL in array WAVEAB (IPX, IPY)
IPX = IDENSX+1 / 2 ?
Set WAVEAB (IPX, IPY) for RHS of pupil equal to WAVEAB (IPX, IPY) for corresponding point in LHS of pupil
NO
7
Set current IPY
Set I=1
Interferogram analysis
YES
IPY = IDENSY ?
YES
IPX = IDENSX ?
NO
8
6
3
8
ISTOP = 1 ?
YES
APRINT (1,IPY) ASTOP 'Stops' surround pupil
NO
ISTOP = 0 ?
YES
APRINT (1 IPY) BLANK Blanks surround pupil
NO
9
Set current IPX
Calculate CODE 3 CODE 4 RAD 34
NO
Was ray vignetted ?
YES
APRINT (IPX, IPY) = ASTAR 'X' within pupil
RAD 34 >1 ?
YES
ISTOP = 1
YES
APRINT (IPX IPY) = ASTOP
9
NO
NO
ISTOP = 0
YES
APRINT (IPX IPY) = BLANK
9
NO
10

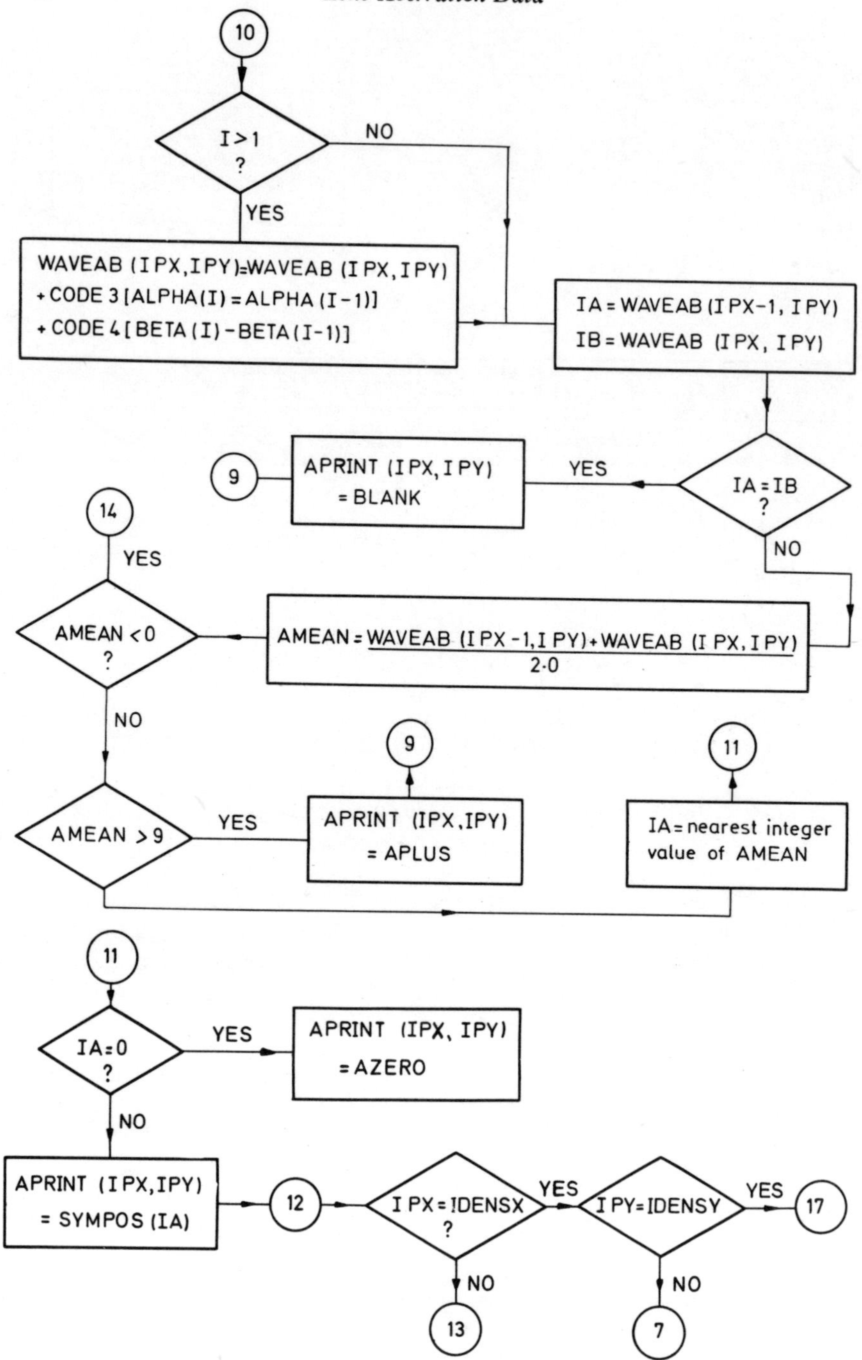

10
I > 1 ?
NO
YES
WAVEAB (IPX,IPY)=WAVEAB (IPX,IPY)
+ CODE 3 [ALPHA(I) = ALPHA (I−1)]
+ CODE 4 [BETA (I) − BETA (I−1)]
IA = WAVEAB (IPX−1, IPY)
IB = WAVEAB (IPX, IPY)
9
APRINT (IPX,IPY) = BLANK
YES
IA = IB ?
NO
14
YES
AMEAN < 0 ?
AMEAN = WAVEAB (IPX −1,IPY)+WAVEAB (IPX,IPY) / 2·0
NO
9
11
AMEAN > 9
YES
APRINT (IPX,IPY) = APLUS
IA = nearest integer value of AMEAN
11
IA = 0 ?
YES
APRINT (IPX, IPY) = AZERO
NO
APRINT (IPX,IPY) = SYMPOS (IA)
12
IPX = IDENSX ?
YES
IPY = IDENSY
YES
17
NO
13
NO
7

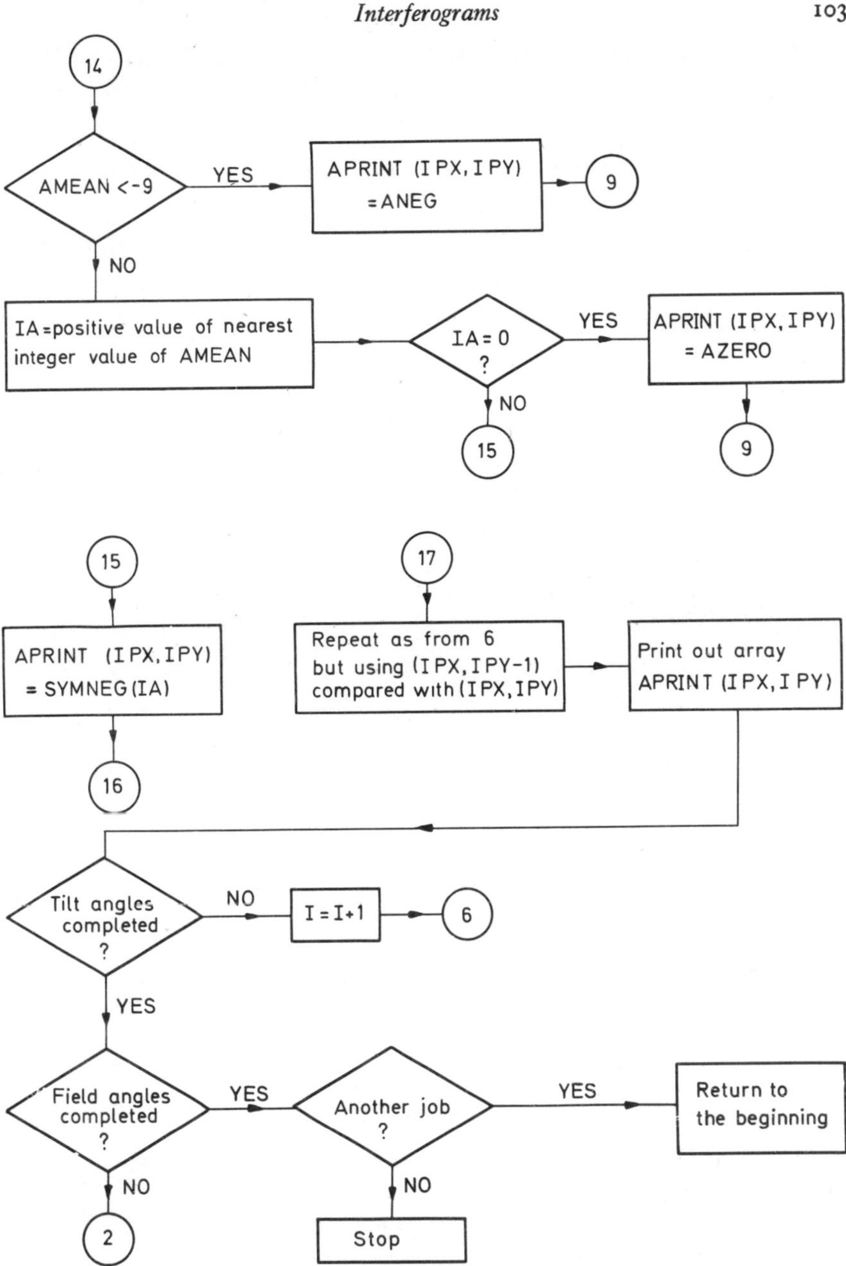

values in excess of nine wavelengths, a minus sign is stored. Finally, the print character array is printed in one operation. Now, it can be seen that if the fringe contour is almost tangential to the horizontal elements in the printing array, large gaps will be left between printed characters and it follows a double scan is necessary to give a reasonable approximation to the contour. If required, a smooth curve can be fitted to the line of printed characters.

In order to include the effect of tilt of the reference surface the wavefront aberration array is re-scanned and a component of tilt added to each element appropriate to its position in the pupil. In fact, an amount for the current tilt is added and an amount for the previous tilt subtracted. The first tilt must therefore be defined as zero. There is the option of a tilt α, about the meridian section or a tilt β about the sagittal section.

$$W_{i+1} \rightarrow W_i + \text{CODE } 3\,(\alpha_{i+1} - \alpha_i) + \text{CODE } 4\,(\beta_{i+1} - \beta_i)$$

4.3. APPLICATION OF THE PROGRAM

The program is intended to provide a means of displaying the aberrations of the systems as contours of wavefront aberration so that the three-dimensional characteristics of the wavefront can be visualized. This is a very elegant method of presentation from the point of view of the designer as the wavefront contains all the information of the image. It does, however, require considerable experience of interpretation to make full use of this information. Kingslake (1926) has treated the case of interferometer patterns in the presence of primary aberration. He gives two methods for the calculation of the contours from a knowledge of the primary aberration coefficients of the power series expansion of wavefront aberration. The theoretical interferograms are also compared with measured interferograms of telescope objectives with known amounts of pure primary aberrations.

This technique was taken up by Coleman, Clark and Coleman (1947) and extended to include simple mixtures of aberrations and extra-axial images. The interferograms reproduced in these papers are most valuable as a means of familiarizing oneself with he effect of aberrations on an interferogram. Coleman *et al.* also studied the possibility of evaluating the quality of telescope systems on the basis of an interferogram. They found that in the presence of mixtures of aberrations it became very difficult to analysis the pattern for individual component aberrations on the basis of shape and number of fringes. Consequently, they suggested a criterion of quality based on the diameter of the largest circle that could be inscribed within the area of a 'fluffed-out' fringe, that is, when the tilt is adjusted so that one fringe encloses an area covering most of the pupil. This criterion was said to correspond to the best visual focus obtained with the telescope.

Hopkins (1955b) was able to show that measuring the integrated intensity of a point source diffraction image in a wavefront shearing interferometer gives a more theoretically sound measure of the quality of a system. This method expressed as the OTF has now largely replaced the criterion due to Coleman *et al.*, although in fact the method of measurement usually differs from the one suggested by Hopkins.

The two main applications of the program are in design and testing. In the former case, the interferograms will permit the optical designer to decide whether the design meets the specification in terms of, say, the Rayleigh limit. In the second, the testing engineer can compare the interferogram of the design objective with that of the manufactured objective. A simple tolerance may be specified on the manufactured objective. For example, it may be stated that the interferogram of manufactured objective must not depart from the theoretical interferogram by more than a quarter of a fringe, or whatever quantity is appropriate to the objective. Comparatively little skill of interpretation is then required in the quality control of the manufactured objectives.

The program is also capable of showing the shape of the vignetted pupil for a given set of surface diameters. If the diameters are not specified the un-vignetted pupil is produced.

The interferograms for the double Gauss example are shown in Plates 4-6.

5

Appraisal of the Methods of Analysis

5.1 THE ANALYSIS AND PRESENTATION OF ABERRATION IN RELATION TO LENS DESIGN

5.1.1. *Ray generation and ray tracing*

From even a brief scan of the earlier chapters it is clear that ray tracing is the principal method of determining the data required for aberration analysis. This fact is all the more remarkable since the geometrical ray is a conceptual aid defining the loci of the path of energy transmitted by the lens system, and cannot itself be physically isolated. In principle, the process of ray tracing is quite simple, being the successive application of Snell's law of refraction. Even diffraction methods rely on ray tracing to obtain the wavefront aberration which is needed to determine the wavefront aberration polynominal. No analytical process has successfully challenged ray tracing—this being essentially a trigometrical process—as a means of determining the aberration present in a lens system. This dependence on ray tracing probably stems from the initially surprising fact that even the simplest lens system has a large number of characteristic parameters such as curvatures, separations of surfaces, refractive indices and Gaussian parameters (e.g. numerical aperture and stop position). Any effort to formulate the aberration in an analytical fashion leads to intractable mathematics and consequently flounders. Hence, we may expect to continue to rely on ray tracing as the major source of aberration data.

The exercise of developing methods of automatic ray generation is an example of the right use of electronic computers for scientific purposes. The good program will aim at minimizing the input and output and, at the same time, exploiting the computer's capacity for extensive calculation. In practice we may incorporate, as part of ray generation, a routine which sets up the Gaussian optics of the lens system. The speed achieved in ray tracing permits us to employ iterative procedures. If, for example, the stop position within the lens system is specified the program may trace outwards into the object space, thereby determining the principal ray data in the object space for a given field angle. It will be necessary to take an arbitrary principal ray angle at the stop and iterate until the angle in the object space corresponds to the required field angle. Similarly, the program may scale the system to provide a particular magnification for a given object-to-image separation. When this type of procedure is coupled with the aperture-filling ray tracing routine, we have a convenient means of undertaking comprehensive image analysis.

5.1.2 *Total ray aberration*

At one time, image analysis was generally made in terms of longitudinal ray aberration. Conrady preferred this form of analysis because it followed simply from trigonometrical ray tracing. In addition, the effect of a change in focal plane is merely to move the origin of the plot without otherwise changing the shape or orientation of the curve. On the whole, longitudinal aberration is satisfactory for making an appraisal of image quality produced by a lens, provided the system has a long front conjugate and a small field angle. The long front conjugate avoids the difficulty experienced with beams of low convergence when the magnitude of the aberration tends to infinity. The restriction to a small field angle ensures one avoids forming an erroneous impression of performance through ignoring skewing aberration. The aberrations of the ray bundle including skew rays may be treated as a longitudinal error if diapoint analysis is used. The difficulty of errors tending to infinity is overcome, for example in oculars, by calculating reciprocal errors. As with longitudinal aberration, the plot of diapoints is independent of the choice of focal plane.

The conventional classification of aberration is difficult to apply to the graphical display of a diapoint cluster. An alternative, given by Hilderbrand (1967), classifies the errors as (i) aperture errors, with the diapoints falling on a straight line, (ii) aperture errors and errors of asymmetry, with the diapoints falling on a curve, (iii) aperture errors, errors of asymmetry and deformation error, with the diapoints forming a surface. Whilst these errors make an interesting analysis, it is not yet clear how they can be related to the information necessary for correcting a system. It would seem that diapoint analysis gives no additional information about the aberration but is useful when taken in conjunction with other forms of display, for it does give information about skew ray aberration.

The lateral or transverse ray aberration is undoubtedly the most useful type of aberration to display. In its simplest form, meridian ray transverse aberration, there is a direct relationship to wavefront aberration and, consequently, inspection of the shape of the curves leads to information about the relative significance of various terms of wavefront aberration. The more general form of transverse aberration is the spot diagram, which gives a clear guide to the geometrical image of a point object. The main value of the spot diagram is its use in assessing the importance of the contribution of skew rays to the general aberration patch. Various methods of labelling the rays according to their origin in the pupil have been proposed and the modified form of spot diagram due to Kinglake (1966) is especially interesting. The spots are transferred back to the pupil and represented as an array of vector diagrams of transverse error. Thus they clearly indicate the regions of the pupil which transmit rays subject to serious aberration.

5.1.3 *Specialized forms of ray aberration*

The user of an optical system is generally not concerned with the form of the image defect. The relative content of coma and astigmatism are not so important as the ability of the lens to form a faithful reproduction of the original or to transmit some information satisfactorily. The lens designer, on the other hand, while endeavouring to satisfy the user's demands, will need to know why a particular design fails to meet the specification and will need information to guide remedial action. Approximate knowledge of the relative magnitude of various terms of the wavefront aberration function or groups of similar terms is generally sufficient for most purposes. The techniques of analysis available through evaluating OSC′, sagittal and tangential focal errors, and distortion are convenient because they involve comparatively simple calculations compared with the much more elaborate calculations necessary for obtaining the coefficients of the wave aberration function.

As an example of the value of such methods, let us consider a common design problem, that of the achromatic and aplanatic air-spaced doublet. Once the chromatic condition is satisfied it is only possible to correct two other aberrations. The problem is to determine the bendings of the shapes of the two components in order to correct the spherical aberration and coma. We need only trace a few axial meridian rays to calculate the transverse spherical aberration and the associated OSC′. From these we calculated the bending differentials and hence develop two linear simultaneous equations for the spherical aberration and OSC′. The solution to these equations gives a bending for each component. The lens with new shapes is ray traced and should give an improvement in correction of at least an order of magnitude. Resolving and ray tracing to obtain successively better solutions provides the corrected design in a few iterative steps. Within the limits of field set by astigmatism and field curvature the resulting design will be found to be substantially free from coma when subjected to careful analysis.

A cemented doublet may be designed along similar lines but correction is even further restricted. The aplanatic condition can only be satisfied if a strictly limited range of achromatic glass pairs are used. It becomes difficult to develop a systematic method because the design parameters are not independent. The initial choice of glass pairs may be made on the basis of thin lens calculation and the assessment made on the basis of meridian ray tracing and OSC′ calculation.

Both these design problems are well suited to the simple method of analysis first developed by Conrady.

5.1.4 *Geometrical wavefront aberration and interferograms*

It has long been recognized by practical opticians that the asphericities of the image-forming wavefronts determine the quality of the image. Twyman was an advocate of interferometric testing of the wavefront and introduced interferometers with lens-testing facilities into the workshops of Adam Hilger Ltd.

In design calculations the wavefront aberrations can be displayed graphically as sections through the wavefront. From two orthogonal sections for each field angle an estimate can be made of the magnitude and form of the defects present in a lens. Alternatively, interferometic plots similar to those shown in Plates 4–6 may be prepared, but such plots are really intended for final testing purposes.

The basis for assessment of geometrical wavefront aberration can be made at one of several levels. It may be sufficient to use a tolerance such as the Rayleigh limit and simply to inspect the error curve to discover whether the wavefront will fit between two concentric spheres centred on the image point and separated by a quarter wavelength. The Maréchal tolerances based on the Strehl intensity ratio may be applied where the magnitude of the coefficients of wavefront aberration are known. Even more refined analysis is possible if the wavefront aberration is merely used as an intermediate stage in computing the OTF. For this purpose the first step is ray tracing to find the wavefront aberration at a few points in the pupil, then a polynominal is fitted to the aberration data, and finally the transfer integral is evaluated. The Hopkins OTF tolerances may be appropriate for the aberration analysis provided the lens falls within the range of validity of the tolerances.

Wavefront aberration is particularly useful because it represents a link between purely geometrical ray errors, such as transverse aberration, and the more rigorous treatment of image formation by physical optics as described by the OTF. The wavefront aberration is a little more troublesome and costly to compute than the transverse aberration, but is less so than the full diffraction-based OTF calculation or even its geometrical approximation. The Conrady (*d-D*) approximation provides a useful method of saving cost when wavefront aberration is required at more than one wavelength.

Probably two of the most important considerations in selecting a suitable form of analysis are the form of the object and the magnitude of the residual aberration inherent in a particular class of lens design. The dependence on the form of the object means that a lens intended for imaging a point object will be treated differently to one for wide-angle reproduction. The first will require a very high Strehl ratio whereas the second will demand freedom from distortion and even illumination. The magnitude of the aberration is important since purely geometric analysis is satisfactory only for comparatively large amounts of aberration. As the performance of a lens approaches that of a theoretically perfect lens, it becomes necessary to consider image formation as a diffraction phenomenon.

Finally, a cautionary note on attempting to economize by limiting the scope of the analysis may not be out of place. Whenever the aperture field product reaches the magnitude of the double Gauss example, then care must be taken to examine the skew ray aberration. It is not sufficient to concentrate on meridian section rays. The magnitude of the sagittal section rays in the double Gauss example is by no means insignificant. In this case the predominant aberration is astigmatism and the sagittal focal errors warn us of its presence. Had the skew aberration consisted of oblique spherical aberration, a serious defect might have been missed.

5.2 ABERRATION TOLERANCES IN THE SPECIFICATION OF THE LENS SYSTEM

Before starting his work proper a lens designer must first establish the target specification. Generally, the potential user of an optical system is unable to present his requirements in terms suitable for the designer's work, for example, in terms of transverse aberration. Probably the user will be precise about the function of the lens and will be able to supply a Gaussian layout. The performance of the lens, on the other hand, may well be in terms of detector performance. In a television system, for example, the spatial frequency response of the Vidicon will be known. Alternatively, the performance may reflect the basic accuracy claimed in the instrument, for example, a specified angular sensitivity in a theodolite. Hense, in a specification for a designer the aberration tolerance is an invaluable quantity.

Initially the tolerances divide the lens systems into two groups, the high-performance diffraction-limited systems and the less highly corrected systems.

Systems designed for handling images of very small objects or images with very great accuracy tend to require diffraction-limited lenses. As the physical dimensions for these systems are concerned with the magnitude of the wavelength of light employed, the aberration is almost fully corrected. The tolerances determine the permitted magnitude of the residual aberration and the Strehl intensity ratio is suitable for describing the quality of point imagery. Developed from the Strehl intensity ratio, the Maréchal tolerances provide a means of ascribing values to the residual aberration.

The OTF can also be used to describe the expected performance of a system completely. The advantage of the OTF method is that it deals with extended objects, but there are disadvantages—the cost of computation and the difficulty of relating line frequency performance to the imaging of pictorial objects or scenes found in nature.

The vast majority of lenses manufactured are far from being diffraction-limited and must be toleranced differently. A frequently used but simple method is to specify the diameter of the geometrical aberration patch in various parts of the field. For this purpose, plots of transverse aberration or spot diagrams are examined to determine whether the lens is within the specification. In some applications, particularly for images of point objects or where point-by-point scanned images are in use, the encircled energy is a valuable technique. The tolerance on performance will probably be given as 80 per cent of the encircled energy within a circle of some specified radius.

The less highly corrected lenses are suitable for specification by OTF performance. The design can be evaluated in terms of spatial frequency response using the geometrical approximation, and consequently the cost factor will be significantly less than it is for the diffraction-limited lens. However, the difficulty of anticipating the spatial frequency content of typical objects remains a problem. One other advantage of an OTF tolerance is the possibility of specifying the response to low spatial frequencies. This is useful for cases where a cut-off frequency far from the diffraction limit may be acceptable, but where a high response is required at low spatial frequencies. The typical example is where electronic image tubes are used for detection. The tolerance condition will be specified as, say, 80 per cent contrast at a particular frequency. Alternatively, values of contrast may be given at a series of optical frequencies representing high, medium and low frequencies.

We have seen that the Hopkins tolerances can be used to specify the permissible amounts of various terms of wavefront aberration. These tolerances are suitable for lenses required to provide a good response up to about one-tenth of the diffraction limit. The tolerance on the aberration is then much wider than that given by Maréchal or Rayleigh tolerances.

5·3. ASSESSING THE QUALITY OF MANUFACTURED LENSES

Every design consideration up to the stage of preparing the detail drawings will have been a compromised balance. Now, in the production stages whilst every effort will be made to reach the nominal values of the design, differences from the nominal values will occur even if no actual errors are made. The tendency will be for the original design balance to be upset since all the parameters will be slightly different from the nominal values and the surfaces will not be perfectly centred, or even spherical, as was assumed in the design calculations. Hence, the lens will exhibit aberration due to these effects in addition to any residual design aberration. An attempt can be made to exert a statistical control over the manufacturing variations by applying, at the design stage, tolerances to dimensions and surface quality.

It is desirable, however, that quality control or inspection should follow manufacture with the aim of ensuring the performance of the complete optical system is held to within some tolerance about the design target. In this respect, the two most useful tests are the interferogram and OTF measurements. In either case, the results are quantitative and can be compared directly with the hyperthetical design system. In principle, the two methods are substantially equivalent, since the OTF may be determined from the integrated intensity of a sheared interferogram, the spatial frequency of the measurement being a function of the degree of shear. In practice the measurements appear to be different because the OTF is expressed in terms of the image plane parameters whereas the interferogram is formed in the pupil of the lens. The residual aberration of a design may be represented by the optical transfer plots discussed in Chapter 3 or the interferogram derived by the techniques discussed in Chapter 4. Thus the inspection department can be provided with a graphical representation which will enable them to distinguish between residual design aberration and aberration from manufacturing defects.

Bibliography

BARAKAT, R., 1961a, *Progress in Optics*, Vol. 1, 67–108, (North Holland, Amsterdam).

BARAKAT, R., 1961b, *J. Opt. Soc. Am.*, **51**, 152–157.

BARAKAT, R., 1962, *J. Opt. Soc. Am.*, **52**, 985–990.

BARAKAT, R., and MOVELLO, M., 1964, *J. Opt. Soc. Am.*, **54**, 235–237.

BORN, M., and WOLF, E., 1964, *Principles of Optics* (Pergamon Press, Oxford).

BRAY, C. P. C., 1965, *J. Opt. Soc. Am.*, **55**, 1136–1138.

BROMILOW, N. S., 1958, *Proc. Phys. Soc.*, *B***71**, 231.

CHANCE–PILKINGTON, 1966, Chance-Pilkington glass catalogue (Pilkington Bros. Ltd, St Asaph, North Wales).

COLEMAN, H. C., CLARK, D. G., and COLEMAN, M. F., 1947, *J. Opt. Soc. Am.*, **37**, 671–685.

CONRADY, A. E., 1929, *Applied Optics and Optical Design*, Vol. 1 (Constable, London).

CONRADY, A. E., 1960, *Applied Optics and Optical Design*, Vol. 2 (Dover, New York).

COX, A., 1964, *A System of Optical Design* (Focal Press, London).

DE, M., 1955, *Proc. Roy. Soc.*, **233A**, 91.

DUMONTET, P., 1955, *Optica Acta*, **2**, 53.

FRANÇON, M., 1966, *Optical Interferometry* (Academic Press, London).

GOODBODY, A. M., 1958, *Proc. Phys. Soc.*, **72**, 411–424.

GOODBODY, A. M., 1960, *Proc. Phys. Soc.*, **75**, 677–688.

HAWKINS, D. G., and LINFOOT, E. H., 1945, *Monthly Notices Roy. Astrom. Soc.*, **105**, 334.

HERZBERGER, M., 1936, *J. Opt. Soc. Am.*, **26**, 197–204.

HERZBERGER, M., 1947, *J. Opt. Soc. Am.*, **37**, 485–493.

HERZBERGER, M., 1957, *J. Opt. Soc. Am.*, **47**, 584–594.

HILDEBRAND, K., 1967, *Untersuchung des Korrektionszustandes von Photo-Objektiven nach einer von M.Herzberger vorgeschlagenen Methode der Analyse der Diapunktsaberrationen* (Wild Heerbrugg AG).

HOPKINS, H. H., 1946, *Proc. Phys. Soc.*, **58**, 685.

HOPKINS, H. H., 1950, *Wave Theory of Aberrations* (Oxford University Press).

HOPKINS, H. H., 1952, *Proc. Phys. Soc.*, **65B**, 934–942.

HOPKINS, H. H., 1953, *Proc. Roy. Soc.*, **A217**, 408–432.

HOPKINS, H. H., 1955a, *Proc. Roy. Soc.*, **A231**, 91–103.

HOPKINS, H. H., 1955b, *Optica Acta*, **2**, 23–29.

HOPKINS, H. H., 1957, *Proc. Phys. Soc.*, **70B**, 447–470.

HOPKINS, H. H., 1966, *Optica Acta*, **13**, 343–369.
HOPKINS, R. E., 1962, *Proceedings Conference on Optical Instruments and Techniques*, p. 65. (Chapman & Hall, London).
HOPKINS, R. E., 1967, *Notes on Image Evaluation*, Summer School 1967, (University of Rochester, Institute of Optics, N. Y.).
KIDGER, M. J., and WYNNE, C. G., 1967, *Appl. Optics*, **6**, 553–563.
KINGSKALE, R., 1926, *Trans. Optical Soc. XXVII*, **2**, 94–105.
KINGSLAKE, R., 1966, *Lens Design with Large Computers*. (Proceedings of a Conference at the Institute of Optics, University of Rochester, N. Y.).
LINFOOT, E. H., 1955, *Recent Advances in Optics* (Oxford University Press).
MARÉCHAL, A., 1947, *Revue d'Optique*, **26**, 257.
MARÉCHAL, A., and FRANÇON, M., 1960, *Diffraction Structure des Images* (Editions de la Revue d'Optic).
MIYAMOTO, K., 1957a, J. Appl. Phys. Japan., **26**, 421.
MIYAMOTO, K., 1957b, *J. Opt. Soc. Am.*, 47, 774–779.
MIYAMOTO, K., 1961, *Progress in Optics*, Vol. 1., 33. (North Holland, Amsterdam).
NIJBOER, B. R. A., 1942, The diffraction theory of aberrations. Thesis, University of Groningen. (Published in part in *Physica*, 1947, **10**, 679; 1947, **13**, 605.)
RAYCES, J. L., 1964, *Optica Acta*, **11**, 85–88.
RAYLEIGH, LORD, 1881, *Phil. Mag.* (5), **11**, 214–218.
RAYLEIGH, Lord, 1879, *Phil. Mag.* (5), **8**, 403.
SHANNON, R. R., 1969, *Proceedings of a Conference of the ICO*. (Oriel Press, Glasgow).
SMITH, T., 1945, *Proc. Phys. Soc.*, **57**, 286.
SOUTHALL, J. P. C., 1913, *Principles and Methods of Geometrical Optics*, 2nd edn. (London).
STRAVROUDIS, O. N., and FEDER, D. P., 1954, *J. Opt. Soc. Am.*, **44**, 163–170.
STEEL, W. H., 1956, *Optica Acta*, **3**, 67.
STEEL, W. H., 1957, *J. Opt. Soc. Am.*, **47**, 405–413.
STEEL, W. H., 1967, *Interferometry* (Cambridge University Press).
WANDERSLEB, F., 1952, *Die Lichtverleilung in der axialen Kanstik eines mit Spharischer Aberration behafteten Objectives* (Berlin).
WELFORD, W. (writing as Weinstein, W.), 1952, *Proc. Phys. Soc.* **65B**, 731.
WELFORD, W. (writing as Weinstein, W.), 1956, Contribution to the Symposium on Optical Design with Digital Computers, Imperial College, Technical Optics Section, London.
WELFORD, W. T., 1963a, *Guide to Instrument Design*, Edited by Smith, A. W., and Thompson, J., (Published for SIRA by Taylor and Francis, London).
WELFORD, W. T., 1963b, *Optica Acta*, **10**, 121–127.
WELFORD, W. T., 1967, *Handbuch der Physik*. **XXIX**, 1.
WOLF, E., 1951, *Proc. Roy. Soc.* (*London*), **A204**, 533.

INDEX